Life's Blueprint

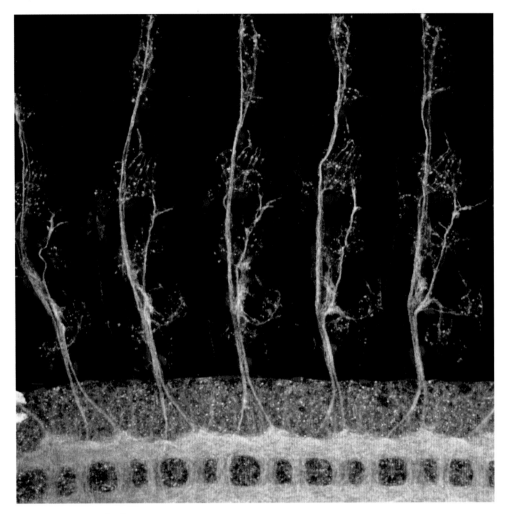

Trees of embryonic neurons

Life's Blueprint

The Science and Art
of Embryo Creation

Benny Shilo

Yale UNIVERSITY PRESS NEW HAVEN AND LONDON

Published with assistance from the Alfred P. Sloan Foundation Public Understanding of Science and Technology Program and from the foundation established in memory of Amasa Stone Mather of the Class of 1907, Yale College.

Yale University Press books may be purchased in quantity for educational, business, or promotional use. For information, please e-mail sales.press@yale.edu (US office) or sales@yaleup.co.uk (UK office).

Printed in China.

Library of Congress Cataloging-in-Publication Data
Shilo, Benny, 1951– author.
Life's blueprint : the science and art of embryo creation / Benny Shilo.
pp ; cm.
Includes bibliographical references and index.
ISBN 978-0-300-19663-4 (clothbound : alk. paper)
I. Alfred P. Sloan Foundation. Public Understanding of Science and Technology Program, sponsoring body. II. Title.
[DNLM: 1. Embryonic Development. 2. Embryonic Structures—cytology. QS 604]
QM601
612.6'4—dc23   2014014622

A catalogue record for this book is available from the British Library.

This paper meets the requirements of ANSI/NISO Z39.48–1992 (Permanence of Paper).

10 9 8 7 6 5 4 3 2 1

Frontispiece: Trees of embryonic neurons. The sensory neurons of the peripheral nervous system of the fruit fly embryo connect to the central nervous system, seen as a ladder at the bottom. They transmit information from external sense organs and internal muscle stretch receptors. The structure of the neurons is repeated accurately at every segment of the embryo. Some of the neural cell nuclei are marked in red, and nerve axons in gray (M. Silies and C. Klämbt, Münster University)

To my parents,

Prof. Moshe Shilo and Dr. Mira Shilo,

who embedded in me the passion for science

and introduced me to the wonders of biology

Contents

Preface

D uring fertilization, the genetic contents of the egg and sperm, each carrying half of the genetic material of the mother and father, respectively, are united in a single cell. This unicellular embryo will give rise to a complete multicellular organism, with its enormous complexity of tissues and multiple types of cells that will appear different from one another and carry out diverse and distinct roles. All the information for this elaborate process is carried within the DNA of that initial single cell that was formed on fertilization, and all cells of the embryo will contain exactly the same genetic information (fig. 1).

The long-standing human desire to understand the instructions and rules of forming an embryo and to unravel "life's blueprint" stems from many sources. First and foremost, we want to understand how we as humans, and all other organisms, build our complex bodies. How do genes shape animals? What are the instructions? How are they carried out in such a reliable and reproducible fashion? Can we use this knowledge to better understand pathological situations in which the genetic program goes awry? And, thinking to the future, can we use the information gained in order to instruct cells to generate tissues and organs that could be used as "spare parts"?

As we set out to explore these fundamental questions, more specific and operational issues come to mind. Are cells preprogrammed by the genome to generate distinct cell types, or do cells have a choice of assuming different fates? How do they decide which fate to choose from the arsenal of tis-

Figure 1. *All cells in our body carry identical genetic information*
Upon fertilization, the nuclei of the sperm and egg join, and the first cell of the embryo is formed. Throughout the numerous divisions that will follow, all descendants will contain the same genetic legacy. This information dictates the shape of the embryo, as indicated by the fact that identical twins carrying the same genetic material are so similar.

Top: Scanning electron micrograph of fertilized human egg (F. Leroy/Science Source); *bottom*: Identical twins (B. Shilo, Rehovot, Israel)

sues they are capable of forming? Does every species have its own rules and logic of embryonic development, or, alternatively, can we identify common themes among disparate species? The scientific discoveries of the past three decades have revolutionized our knowledge and understanding of embryonic development in all multicellular organisms. This book conveys these striking findings and the emerging rules of executing development.

The analysis of embryonic development relies strongly on microscopic images that we obtain and analyze. These images include fluorescent multicolor depictions of different embryonic tissues, scanning electron microscope images, and movies of live tissues and embryos showing dynamic processes in real time. The images are not only highly informative but often also aesthetically striking. Their visual beauty is often central in attracting many researchers to the field.

For me, these visual scientific stimulations are combined with another personal passion. Since my high school years, I have been interested in photography and how we view and depict our surroundings. In my early work, I felt more comfortable photographing still objects and how they interact with light. In the past fifteen years, as a result of frequent trips to foreign destinations, including the countries of the Far East and India, and possibly also as a consequence of a more mature view of the world, I have concentrated on photographing people. I try to portray universal aspects that unite people of all races and ages, as well as the unique features of individuals in their surroundings.

To visually present the emerging concepts of embryonic development in this book, I have assembled scientific images of diverse organisms that display the central concepts of embryogenesis from the laboratories of colleagues and from my own lab. Each of these images is paired with an image from our "macro" world taken by me, highlighting a similar theme and reflecting my personal metaphors. In seeing these juxtaposed images, I hope that readers will be able to understand each paradigm in an intuitive way, letting our everyday experiences and recollections resonate with and inform our comprehension of the scientific image.

It is interesting to note that the inception of the word *cell* as a biological term was formulated by Robert Hooke in 1665 and was based on a visual analogy to the macro world. Hooke looked at the bark of a cork oak using

a microscope he developed. He saw the regular shapes of the nonliving cell walls and noted that they resemble the cells in which monks live. I envisage that through the pictorial analogies, the reader will "feel" like a cell within the developing embryo (fig. 2).

The goal of this book—to present a coherent and simplified description of the complicated process of embryonic development—has forced me to define to myself the true essence of each aspect of life's blueprint. The juxtaposition of metaphors from the human world with these biological processes has also required a continuous examination of each concept in order to crystallize its essence and employ the most appropriate analogy. Finally, my efforts to present a global view of embryonic development have mandated a critical evaluation of the relative importance of each concept from a broader perspective. As a scientist, my work involves delving deeper and deeper into defined biological problems. Stepping back and viewing the entire scope of embryogenesis in a global manner has been both refreshing and illuminating.

◀ *Figure 2. Analogies between cells and the human world*
From the inception of the word by Robert Hooke in 1665, the cell has been defined by analogy to the human world. Hooke coined the word *cell* because he felt that the structures he saw in his microscope resembled monks' cells. In this book I will be making analogies between the environment of cells and the human world. I would like the viewer to "feel" like a cell as it participates in forming the complex structure of the embryo by relating this process to visual and conceptual similarities in the human world.

Top: Drawing of cork oak bark by Robert Hooke in *Micrographia*; *bottom*: Child and cell (B. Shilo, Exploratorium, San Francisco)

Acknowledgments

The exchanges with so many who contributed their unique expertise and points of view have made the creation of this book a pure joy. During my career I have been inspired by my father, Moshe Shilo, who was a pioneer in microbial ecology; my PhD mentor, Giora Simchen; my postdoctoral adviser, Robert A. Weinberg; and the numerous colleagues with whom I have interacted throughout the years. Especially vital have been the exchanges with the students and colleagues in my lab, who also contributed many of the scientific images displayed in this book. I would also like to thank my international scientific colleagues for generously providing scientific images.

I am truly thankful to the Radcliffe Institute for Advanced Study for hosting me as a fellow during the academic year of 2011/2012 and providing the supportive and multidisciplinary environment that fostered the emergence of this book. The late Lindy Hess at the Radcliffe Institute guided me with her expert and personal touch through the initial and critical stages of daring to write my first book; I am deeply saddened that she will not see the finished product. Dr. Judith Vichniac, director of Radcliffe fellowship programs, provided friendship and support.

I am indebted to Michal Rovner, who inspired me to use black and white with a special combination of color in the images to highlight points of interest; my son, Avner Shilo, who contributed continuous advice in image manipulation; my daughter, Anat Shilo, for printing the images; my grandchildren, Tom and Noa Dubnov, for serving as models in photographs; and my wife, Varda Shilo, who provided the initial trigger for the project, as well as continuous inspiration and discussions as it progressed. Nancy Lynch offered thoughtful comments, and Chani Sacharen edited the text insightfully. At Yale University Press, the support of Jean Thomson Black, who supervised the book from the initial proposal stages to the final version, was crucial. I also thank Laura Dooley for the final editing and Samantha Ostrowski for dedicated administrative assistance. Finally, I am indebted to the Alfred P. Sloan Foundation Public Understanding of Science and Technology Program for support in publishing this book.

CHAPTER 1

Introduction: Exploring the Principles of Embryo Creation

I would like to begin by sharing my personal story and how I embarked on a route that charted my academic career. Along this journey, I was extremely fortunate to be exposed to, and contribute to, the burst of findings that provide us today with a deep and comprehensive understanding of embryonic development. The year was 1981, and I was a twenty-nine-year-old postdoc in the lab of Robert A. Weinberg at the MIT Center for Cancer Research in Cambridge, Massachusetts. During that time, the pioneering work of the lab devised ways to identify and isolate single cancer-causing genes, called oncogenes, from the DNA of human tumors of various origins. I was privileged to be part of the group that created this breakthrough.

Our discovery was preceded by the work of many labs researching RNA viruses. These viruses, which are called retroviruses, hijack normal genes in the cells they infect and convert them to oncogenes by altering their structure or regulation. Although we recognized the cellular origin of numerous oncogenes, we still did not understand the role they play within healthy cells. The functions of these cellular ("normal") oncogenes, before they are hijacked and modified by the retroviruses, must represent central junctions of regulation for the cells. This is why subtle changes in their activity can lead to such deleterious outcomes as cancer. But what are these normal functions?

The answer requires a global view of the entire repertoire of living organisms through the lens of evolution. Instead of viewing all organisms in a "snapshot" as they appear today, we need to consider how this incredibly diverse array of living forms came to be. The appearance of life on earth was a rare process that took place once or at most very few times. A key feature of any living entity is an ability to replicate itself. Through natural selection the most suitable progeny was chosen for each environment. Over the billions of years since life first appeared, life forms have diversified and expanded, giving rise to the vast array of bacteria, plant, and animal species living today.

The process of evolution is dynamic. Hence, many organisms that formerly existed are no longer with us, and we can only glimpse their structure from fossil records. On the other hand, in general, the commonalities shared by different organisms today reflect their shared origin. It is likely that the features common among organisms are generally inherited from a joint ancestor rather than invented multiple times independently. One of the best examples of this shared inheritance is the genetic material, DNA. Short for "deoxyribonucleic acid," DNA is a self-replicating material that is present in the cells of all living organisms. Perhaps there could be other chemical solutions to encode genetic information. But all organisms that have descended from the primordial cell in which the structure of DNA first arose are "stuck" with this design. We can thus draw two general intuitive rules. First, when a gene is found in a broad range of organisms, it is likely to play a similar role in those different organisms. Second, the wider the range of organisms in which a given gene is found, the more general or basic its function is likely to be.

That said, in the late 1970s it was not clear whether the genes and biological processes that can be analyzed in simpler organisms, such as the fruit fly or the yeast we use to make bread and wine, bear any relevance to complex biological mechanisms such as the progression of cancer or the development of the brain. Were these intricate processes introduced late in evolution, together with the emergence of complex organisms? Or were these basic biological processes actually "invented" early on in evolution and exist today in a wide range of multicellular organisms to carry out very basic and common biological processes? If the latter were the case, we realized that we could make great strides in our understanding by studying these processes in the simpler organisms and then extrapolating the knowledge to the mechanisms that may operate in complex organisms like humans.

In the late 1970s and early 1980s, scientists had developed techniques to use the genes hijacked by retroviruses to isolate the corresponding normal genes from the genomes of vertebrate species, including the chicken, mouse, rat, cat, and monkey. But no one had as yet explored whether the presence, and hence possible function, of the cellular genes that can give rise to cancer was specific to these vertebrate organisms or extended more broadly within a wider range of multicellular organisms.

My research approach to uncover the normal functions of cancer-causing genes started from a simple notion. If genes similar to oncogenes are also found in the genome of simpler organisms, then we could decipher their normal function by exploring these simpler organisms and then examining whether the concepts we uncovered also applied to humans. Consider a little boy assembling his first pair of Lego interlocking wheels. Once he grasps the concept, he can see that the wheels of a train or in a watch mechanism, though more complex, employ the same basic principle.

How do you look for oncogenes in lower organisms? Today we know the linear order of the building blocks on the DNA (termed the nucleotide sequence) of the entire genome of hundreds of diverse organisms, ranging from bacteria and worms to human beings. But at that time, although the DNA fragments harboring these genes were accessible, their sequences, let alone their functions, were not known. My first step involved collecting DNA samples extracted from diverse multicellular organisms. To assemble this collection, I visited labs in Boston and Cambridge where research was under way on such organisms as sea urchins, nematode worms, flies, and even single-cell organisms such as baker's yeast.

I will now present a crash course in molecular biology as I describe how we looked for genes that were similar in sequence to the cancer-causing genes. DNA contains four types of units called nucleotides; these are termed A, C, G, and T. When we speak about the sequence of DNA we refer to the order in which these units, or bases, are linked to one another to form the genetic information of each organism. To test whether "relatives" of a given oncogene were present in these DNAs, we first needed to divide the very long and continuous strand of DNA into fragments. We did this by using an enzyme that cut the DNA each time it recognized a specific stretch of six nucleotides. Although the DNA was prepared from thousands of identical nuclei, the same types of fragments were generated for every cell. The process was similar to starting a new paragraph in an encyclopedia every time a specific combination of six letters appeared. The same paragraphs would be generated for multiple copies of identical volumes.

We then separated each sample into fragments of discrete sizes by placing them in an agarose gel, which, when combined with an electrical field, separated DNA fragments based on their size and charge. The gel was then overlaid with and blotted onto a special filter paper, and the DNA fragments "stuck" to the filter according to their position on the gel. The filter now contained the entire genetic material of the organisms from which it was extracted. Next came the stage of finding the needle in the haystack— determining whether the DNA of these organisms had sequences related to the mouse or chicken oncogenes.

When you have an electronic version of a book and are looking for a particular sentence you vaguely remember, you ask the computer to find the closest fit to this sentence in the entire book. We employed this same logic in the molecular procedure we used to look for sequences that are related to the cancer-causing genes. The search was based on the most universal and seminal feature of DNA: its capacity to form a double helix. The DNA bases have the chemical feature of forming pairs: T specifically matches with A, and C pairs with G. The second strand in a DNA double helix contains the pairs that match the first strand (fig. 3).

At the time, the DNA fragments harboring the oncogenes were isolated and grown in large amounts by procedures of molecular cloning based on their inclusion within the viruses that hijacked them. The process of molecular cloning is similar, if you will, to photocopying just one page from a thousand-volume encyclopedia over and over again. These are the same proportions between the complexity of the entire mouse or human genome versus a single gene.

Figure 3. *The complementary structure of the DNA double helix* ▶
The stairway and its matching shadow depict the concept underlying the structure of DNA. The bases in one strand of the helix are complementary to the bases of the adjoining strand. This feature enables not only the pairing that generates the double helix but also the ability to replicate the DNA based on the information contained within each strand and the generation of RNA molecules that copy the DNA sequence. This complementarity of bases was used in order to "fish" from the fly genome the sequences that are complementary to human oncogenes (B. Shilo, Copenhagen)

The process of DNA replication, in which one strand of DNA can be used as a template to generate the complementary strand in the double helix, can be carried out in the test tube by the addition of certain enzymes. If one of the four bases is radioactively labeled, then the entire strand of DNA that is generated will also be labeled. If many copies from a short strand of DNA containing an oncogene can be marked, they can then be used as a "probe" to explore the vast array of DNA fragments on the filter in a search for DNA stretches that harbor a similar sequence with which they can pair. Even if the oncogene and the DNA on the filter share only a partial identity (that is, they are relatives but not identical twins), this similarity might suffice to allow them to associate with each other.

The filter was then immersed in a radioactively labeled sample of the gene to be tested. After the filters were cleansed of the excess radioactive oncogene probes, the only remaining probe would be bound to the filter at the spots where it found complementary sequences. This binding was then detected by exposing the filter in a sealed cassette to a sheet of film similar to the film used when our teeth or bones are X-rayed. The radioactive probe excited the film and generated a reaction that is similar to exposure to light. The film was then processed in a darkroom by immersion in solutions of a developer and a fixer.

Once I had developed and fixed the film in the dim orange light of the darkroom, I got my first glimpse of the encounter with the labeled oncogenes. There they were, multiple black stripes indicating the presence of similar sequences! Each one reflected the existence of a distinct DNA fragment that paired with the oncogene probe. In the darkroom I could not make out the precise position of the spots on the filter, and hence determine which of the organisms possessed these sequences. But when I later scrutinized the film over a light box, I found that all the spots corresponded to the DNA extracted from the fruit fly *Drosophila melanogaster*.

The implications of this finding were instantly apparent. If the same genes that cause cancer in higher-level vertebrate organisms are also present in fly DNA, this means that they first appeared more than six hundred million years ago, before the evolutionary divergence of vertebrates and invertebrates. Their ancient origin suggests that these genes perform an essential common function in all multicellular organisms. From an experimental standpoint, our ability to manipulate fruit flies, which have been genetically dissected for more than a hundred years, could tell us what functions these same genes carry out in our own human bodies.

For any scientist, there are a few moments in your life when you know you've uncovered an important phenomenon, never before seen or recognized. Even if you don't completely understand the full significance and implications of your discovery, the findings make an immediate impact. These "moments of discovery" are what we scientists live for. I like to think about such moments as analogous in their nature, though not necessarily in scope, to the discovery of the tomb of Tutankhamen on November 26, 1922. The steps leading to the tomb had been cleared, the dignitaries and sponsors of the excavations were assembled outside, and the British archaeologist Howard Carter opened a small hole in the mud wall containing the original royal stamps. Releasing a gust of air that had been sealed up for more than three millennia, he inserted a candle through the hole. Gradually, as his eyes grew accustomed to the dim light, Carter was able to pick out "strange animals, statues and gold, everywhere the glint of gold." His eye was the first to see what would later become a cornerstone in our perception of human civilization.

In the summer of 1981, after having discovered the normal forms of vertebrate cancer-causing genes in fruit flies, I established my lab at the Weizmann Institute in Israel. I spent the next thirty years investigating the role of these genes during the normal life cycle of the fly. This approach can be regarded as "reverse genetics" in the sense that one starts with a known gene and tries to establish its role by generating mutations in this gene and investigating the consequences. In such an approach, you do not know beforehand what biological functions will be unraveled; you simply "trust your gene," analyze it with the toolkit of technologies at your disposal, and hope to uncover the role it plays in central biological processes. Following this approach, I started my journey with genes that cause cancer in vertebrates and ended up studying the central mechanisms by which cells communicate in forming the fruit fly embryo.

A complementary approach, at that time more common, is to carry out "forward genetics." In other words, one defines the biological process of interest and seeks mutations in genes that will disrupt this process. In this approach, the biological process being interrogated is well defined, but the nature of the genes that will be uncovered is not known. The Nobel Prize laureates Christiane Nüsslein-Volhard and Eric Wieschaus used forward genetics to uncover the genes that underlie embryonic patterning in the fruit fly embryo in the late 1970s. Their work revealed most of the genes needed to pattern the fruit fly embryo and triggered a burst of research activity to isolate these genes and identify their molecular nature. When the genes were eventually analyzed at the molecular level, many turned out to correspond to the same genes studied in my reverse genetics approach that had its origin in vertebrate cancer-causing genes. By the late 1980s, the two approaches, which had begun at opposite starting points and used different methodologies, converged.

In the years since then, scores of labs showed that the genes uncovered by these two approaches encoded elements in the communication signals by which cells "talk" to each other during embryonic development to generate the elaborate pattern of the embryo. The presence of these genes in all multicellular organisms indicated that such communication processes might be universal. In fact, these genes embody the essence of multicellularity. The elaborate process of embryonic development is predictable and nonchaotic because the behavior of cells is guided by distinct "rules" that are embedded within the DNA. And when these rules are followed correctly, a pattern emerges.

Unraveling the mechanisms by which cells communicate at the molecular level and the discovery that these modules are common to all multicellular organisms was a breakthrough in our knowledge of life's blueprint. It also converted modern biology into a more universal discipline, since we now know that seemingly disparate processes in different organisms are actually controlled by the same genes and cell communication pathways.

CHAPTER 2

It's All in the Genes

When we consider all the different cells in our body, we can't help but be amazed by how diverse they are. Each cell type is configured to fulfill its ultimate function in the body. Nerve cells form long extensions that can be up to a meter in length and conduct electrical currents at high speed from one end of the cell to the other. Muscle cells form repeated units of molecular motors that move along tracks to carry out muscle contractions. In the muscles, thousands of cells fuse together to generate giant continuous cells that can extend to a meter in length. Red blood cells play a completely different role, carrying the hemoglobin protein that helps bind oxygen in the lungs and then releasing oxygen within the target tissues. Where does the body store the instructions for generating this diversity, and how are these instructions transformed to actual cellular structures? The answer lies in our DNA.

The nucleus of each cell harbors DNA, which plays two fundamental roles: it carries the information that allows each cell type to produce the components that build up that cell, and it stores all the information for constructing the remarkable diversity of cell types encompassed in the whole organism. This information is faithfully duplicated during cell division. It is also transmitted from one generation to the next after a sperm fertilizes an egg.

Four building blocks, called bases, make up DNA; these are abbreviated A, C, G, and T (for adenine, cytosine, guanine, and thymidine, respectively).

The bases join one another to form extremely long strings. The DNA within a single human cell contains three billion bases, which, if spread out, would form a string a meter long. The genetic material in each cell is divided and packed into units called chromosomes, each carrying a different portion of the genetic material. Every human cell contains forty-six chromosomes.

The DNA is typically found in the shape of a double helix, meaning that each strand of DNA is paired with its parallel strand. As I mentioned in Chapter 1, the pairing of DNA strands is the essential underlying feature behind cells' ability to replicate their DNA. There are preferred sets of attraction between the DNA bases: T pairs with A on the opposite strand, and G pairs with C. So when a cell replicates its DNA, the two paired strands separate from each other, and each strand contains the information to form the missing complementary strand. In a way, each strand serves as a "mold." This complicated process does not take place at once but rather proceeds in a progressive manner, in that the two DNA strands are gradually separated and each of the strands is replicated after the separation. In other words, on the template of each strand, free bases pair and assemble to form the matching strand.

The complementary nature of the DNA strands is also the secret behind the cell's ability to convert the information embedded within the DNA into actual functional cellular components. These are the cell's proteins, each carrying out a unique function. Proteins are also composed of repeated units, but the nature of their building blocks is different. How is the information embedded within the DNA used to form proteins, and what is the unit within the very long DNA strands that corresponds to each protein being made?

As stated, the chromosomes are packaging units, and each one contains a large chunk of the genetic material. But the universal functional unit within the DNA that carries the information required for making a protein is much smaller, and it is called a gene. It is important to note that although the nucleus in each human cell contains three billion DNA bases, only a small fraction of those bases, about 3 percent, constitute genes. It is estimated that even very complex organisms, such as humans or mice, have just under twenty-five thousand genes. Picture the DNA as a desert with very sporadic oases; these sites correspond to the protein-coding regions.

So what is all that "extra stuff" in the DNA, and does it have a purpose?

Some parts of it carry instructions on where and when to create proteins. Other parts have more global structural roles, such as separating genes or contributing to the overall function of the chromosome. And some of it represents excess baggage accumulated through evolution—foreign genetic elements that were incorporated into the DNA and remained.

But back to the genes. The two-step process of making proteins, called transcription and translation, involves individual genes on the DNA. First, the information encoded by the DNA that defines a gene is copied into RNA (short for "ribonucleic acid") molecules. The RNA consists of units that are not only very similar to the DNA units but can also pair with them. The process of copying the information of the gene on the DNA into the RNA (transcription) is therefore very similar to that of DNA replication. In transcription, the gene uses the sequence of DNA bases as a template. A single gene can produce thousands of RNA copies.

I have stated that the DNA is stored in a defined cell compartment called the nucleus. The cellular machinery for making the proteins, however, is located outside the nucleus, in a compartment called the cytoplasm. After an RNA molecule, typically several thousand bases long, is produced in the nucleus, it is exported to the cytoplasm, where it encounters the ribosome and undergoes the second step of making proteins, translation.

The ribosome is a giant molecular machine that catalyzes the translation of the information carried by the RNA sequence into the language of proteins. The building blocks of proteins are amino acids, of which there are twenty types. To construct proteins, the ribosome translates the linear code of the RNA into a linear code of amino acids; these become firmly associated with each other like the cars of a train. The linear order of the amino acids in each protein reflects the information originally carried by the gene on the DNA and converted through an RNA intermediate to the protein.

The resulting proteins are the molecular "machines" that carry out the diverse array of cellular functions, but they cannot carry out their task as linear strings of amino acids. Each protein thus undergoes an elaborate folding process to generate a three-dimensional structure that is dictated by its primary sequence. If you hold a strand of wool and fold it up, you will form a different kind of tangled structure each time. In contrast, every time a protein string of amino acids folds, it eventually creates the same three-dimensional structure, because the "rules" for making the structure are embedded within its linear order. Those rules in turn rely on multiple interactions among the protein's amino acids. As a comparison, think of a creating a seating arrangement at a dinner table in which you try to optimize the arrangement and have each diner sit next to the person he or she likes best. Because so many interactions define the final structure of the protein, it is extremely difficult, if not impossible, to predict the structure the protein will assume based on its sequence, even using the most sophisticated computational tools (fig. 4).

Information embedded in the DNA is thus transmitted and used to generate a protein with a defined sequence, using RNA as an intermediary. But how does this process explain the molecular basis for the differences between a nerve cell and a muscle cell? Every cell in the human body carries the same genetic information that was contributed by the DNA of the sperm and egg upon fertilization. But a given cell does not use all of that information; rather, it employs only selective parts that are suitable for its biological role. As a comparison, think of an encyclopedia, in which different pages are accessed each time depending on which topics are sought (fig. 5).

The selective use of the information stored in the DNA by a particular cell type means that only some of the genes are used as templates to produce the RNA that will serve as a basis for making the corresponding proteins. The other genes remain silent and are not used within the context of that cell type. This is the essence of how one cell type differs from another. There is an elaborate regulatory process by which genes are selected to cre-

Figure 4. *The structure of a protein facilitates its function* ▶
The linear information encoded by a gene is transferred from the DNA through an RNA intermediate and is used as a template for production of proteins. Proteins consist of twenty different building blocks termed amino acids. The composition and order of the amino acids (dictated by the sequence of the gene) determine how the protein will reproducibly fold in three-dimensional space. The final structure of the protein facilitates its function by allowing the protein to pair with proteins or other molecules.

Top: Structure of the bacterial protein alcohol dehydrogenase, which initiates the breakdown of alcohol molecules. Four identical copies of the protein (marked in different colors) associate with one another to generate the higher order structure of the enzyme (Y. Burstein, Weizmann Institute); *bottom:* Salt and pepper containers displaying a reproducible and complementary three-dimensional structure (B. Shilo, Mumok, Vienna)

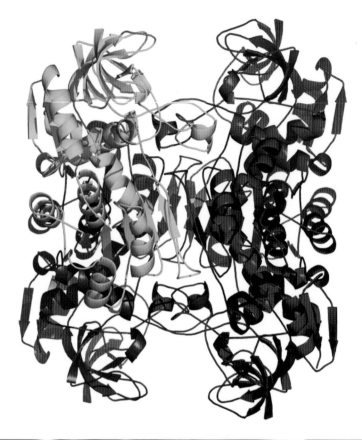

ate RNA. The regulation determines not only which genes will produce RNA but also how many RNA copies each gene will produce and in which order the genes will be used. Just like a cooking recipe, the combination of ingredients, their amount, and the order in which they are added are critical for the ultimate success.

How is the elaborate regulation of RNA synthesis controlled? The machinery for producing RNA chains involves a complex of multiple proteins that slide along the DNA and copy the information of the DNA template to RNA. This apparatus is capable of producing RNA from all genes. So why aren't all genes in the genome active all the time? It turns out that the limiting factor in the production of RNA is how the cell marks those specific genes that will be used to produce RNA. By selectively controlling the association of the copying machinery with some genes and not others, each cell determines the profile of RNA molecules that it will produce.

Now we can define in molecular terms why a nerve cell is so different from a muscle cell. Although they carry identical genetic information on the DNA, each cell type expresses a different set of genes and hence displays a distinct array of proteins that define its structure and function. Only a limited amount of the DNA constitutes the genes and is thus carrying the information for making the proteins. Some of the DNA that is not involved in coding the protein sequence serves to regulate the expression. Gene expression is the process by which the information on the DNA is copied into RNA molecules. The sequences around the gene that are not copied to RNA serve as docking sites for proteins that attach to the DNA. By recruiting a particular set of pro-

◀ Figure 5. Why cells look different from one another

If all cells carry identical genetic information, why do they look so different? For example, a nerve cell is very distinct from a muscle cell. Each cell type uses a different portion of its genetic material as the template for making RNA and proteins. The distinct proteins that each cell type makes reflect and manifest its function.

Top: Scanning electron micrograph of a mouse muscle cell, showing the repeated units called sarcomeres, which carry out muscle contraction (Y. Gruenbaum-Cohen and B. Shilo, Weizmann Institute); bottom: Card filing cabinet in which only some of the information is used at any one time (B. Shilo, Schlesinger Library, Radcliffe Institute for Advanced Study, Cambridge, Massachusetts)

Figure 6. *Coordinating the generation of different cell types*
Although each cell type adopts a distinct shape and function, this process does not happen in isolation. As cells assume different roles, they must be tightly coordinated with the other cells. This enables the creation of such elaborate structures as the mouse limb, which consists of bones, blood vessels, nerves (blue), muscles (red), and tendons (green) (A. K. Lewis, G. Kardon, Y. Wan, and R. Schweitzer, University of Utah and Shriners Research Center, Portland, Oregon)

teins to a given gene, it becomes possible to attract the universal complex of proteins that constitutes the machinery for making the RNA molecules on the basis of the DNA template. Think of a long queue of people standing in line to buy tickets to a movie. As you reach the line you may recognize someone, and you will be immediately attracted to approach the person you know.

How can a population of identical cells that originated from the same fertilized egg generate the intricate structure of the embryo that contains billions of cells and thousands of cell types, each expressing a different set of genes (fig. 6)? In other words, how is the information embedded within the DNA converted into the elaborate structure of the embryo, and how do genes dictate this pattern?

Even at first glance, it is obvious that there is a genetic basis for pattern-

Figure 7. *Embryonic development is a concert without a conductor*

In a concert, the conductor's score shows all the notes that will be played or sung by each musician. Every individual member has the score of all the notes he or she will sing or play. The complete plan is laid out from the beginning. The conductor harbors all the information to coordinate the performance, and the end result is predictable.

It is interesting to compare and contrast this example with embryonic development, where the final result is also reproducible and written down, somehow, in the genome of the cells. But there is no central coordinator who "knows it all." At different stages the cells receive inputs from their neighbors on what to do, but this information relates only to that specific point in time and place, without any clues regarding long-term future actions. Eventually, the local information that is processed individually by each cell gives rise to the elaborate and reproducible shape of the whole embryo, which is infinitely more complex than a Beethoven symphony (B. Shilo, Israel Philharmonic Orchestra and Gary Bertini Israeli Choir)

ing the embryo. Within a given species, all embryos are similar, at least at the global level. The striking degree of similarity between identical twins, which share identical genetic information, is a powerful demonstration of the genetic basis for patterning. Clearly, most of the pattern that will be generated in the embryo is embedded within the genes and the DNA sequences that control them. But how does this happen?

Broadly speaking, and based on our knowledge of organizations within our society, there are two extreme types of management. According to one model, there is a senior executive who has the plans for the final product. This executive organizes and manages groups of workers who will fulfill their distinct tasks. For example, to build an airplane, one group will construct the wings, another will assemble the engines, and so forth. From the outset, this system must have a detailed plan for the completed product. That plan will include how the workers will obtain the parts to be assembled and the established hierarchy of managers who will coordinate and micromanage the activities of the specific task forces. Continuing with our analogy to human interactions, an alternative method would be to have a group of people who are equivalent to one another. The people would establish a defined outcome, setting some general rules for the dynamics of interactions among group members. Although this approach could be an excellent setting for a group of improvisational theater actors, it would not be effective for assembling an airplane.

Is embryonic development more like the assembly of an airplane or the operation of an improvisational theater? An embryo is much more complex than an airplane. But the final plan for the embryo is not laid out from the outset. The parts that will be assembled are not prefabricated, and there is no high-level executive cell who will explicitly instruct each of the other cells which tasks they have to fulfill. In contrast to an orchestra, the elaborate process of embryogenesis progresses without a designated conductor who oversees the entire succession (fig. 7). The shared and identical information that is contained within the DNA does not specify directly the shape of the final product; rather, it writes the rules that cells must follow to reach the desired end point. These rules are spelled out accurately and clearly enough to generate a reproducible pattern. Our challenge is therefore to decipher these instructions. How does a population of identical cells follow such rules to generate the complex and reproducible structure of an embryo?

CHAPTER 3

How an Embryonic Cell "Decides" Its Future Destiny

From Aristotle's time more than two thousand years ago until the recent past, people have believed that embryonic development is deter-
minative and that the final structure of the organism is defined at the
outset—even before fertilization. Embryonic development, it was thought,
simply allows the growth of this structure. This theory of preformationism
claimed that the sperm or egg contains a complete preformed individual,
or "homunculus" (little man). In the first microscopic images of sperm obtained at the end of the seventeenth century, researchers such as Nicolaas
Hartsoeker even drew the little man they were sure they could see within
the sperm. Yet because this theory preceded the birth of modern genetics,
it could not explain how the final shape of the organism showed influences
from both parents.

At the end of the nineteenth century and beginning of the twentieth century, experimental embryologists focused on the question of whether every
cell has a fixed and predetermined destiny or whether small parts of the embryo can re-create the whole embryo. They employed various experimental
tools to follow the developmental fate of parts of embryos that were generated in different ways. For example, after taking a two-celled sea urchin embryo and killing one of the cells with a hot needle, researchers tried to discover whether the remaining cell could create a whole embryo or just the half
embryo it was programmed to form. Alternatively, they separated individual
cells from the whole embryo and tested them for their capacity to form com-

plete embryos. The researchers obtained conflicting results. In some cases, the whole embryo could not be generated, whereas in others complete, though smaller, embryos were formed. Hans Spemann was able to split the two cells in the "two-cell" stage of an amphibian embryo with a fine hair from his baby daughter's head. He found that if the separation was implemented in a particular plane, whole embryos could be generated from a single cell.

These experiments indicated that the preformation theory was wrong and that smaller parts of the embryo could be reprogrammed to form a whole embryo. In 1924, Hilde Mangold, a PhD student in Spemann's laboratory, performed the key experiment that heralded the birth of modern developmental biology. Mangold surgically removed a group of cells from a newt embryo; these cells originally resided in the area that would form the head and trunk of the tadpole. When she transplanted this lump of cells to the flank of another embryo, the transplanted cells became part of the host. To distinguish between the donor and recipient cells, Mangold marked the embryo host cells with pigmentation and used albino cells for the transplanted ones.

This surgical manipulation was crude, and in fact it was successful in only a few of more than ninety transplantations. But the positive results were so stunning that they formed the basis for our current view of developmental biology. This manipulation induced the cells of the host to form a Siamese twin with duplicated head and trunk structures. In fact, the duplication was so accurate that one could not tell the original head from the new one (fig. 8). At face value you might say that the result was not surprising, because the

Figure 8. Cells instruct their neighbors during development ▶

In the classic 1924 experiment by Spemann and Mangold, cells that will give rise to the head and trunk of one newt embryo were transplanted into a host embryo. This manipulation induced the cells of the host to form a Siamese twin. The staggering implication was that cell fates in development are not predetermined. Rather, cells communicate by transmitting and receiving instructions from their neighbors. Therefore, cells encounter multiple junctions in the course of embryonic development, where they have to make decisions that will impinge on their fate.

Top: Embryonic transplantation experiment (H. Kuroda and E. De Robertis, University of California, Los Angeles, and Howard Hughes Medical Institute); *bottom:* Diverging roads in New Hampshire during autumn. How would you choose which one to follow? (B. Shilo)

transplanted cells that were destined to form a head indeed fulfilled their goal at the foreign location. Because the transplanted cells could be distinguished from the host cells, however, the fascinating conclusion was that the host cells formed most of the duplicated head structures. In other words, the transplanted cells "induced" a new fate in the neighboring recipient cells and functioned as an "organizer" for their surroundings. This phenomenon was subsequently named the Spemann organizer.

The term *induction* that Spemann coined embodies the essence of how we currently understand embryonic development. The idea is that cells of one kind influence their neighbors and guide the type of tissue they will produce. Cells communicate with their neighbors and send instructions that alter the behavior of these cells. The cells sending the signal "talk," while the ones receiving it can "listen" and respond accordingly. At the outset, the cells have the potential to go down multiple developmental paths. It is the signals they send and receive that determine their final pattern of differentiation. In other words, animal development results from a succession of cell-to-cell inductions, in which groups of cells direct the differentiation of their neighbors. To understand the processes, we must decipher the choices that cells make, based on the information they receive (see fig. 8).

This new understanding focuses our attention on a fresh set of problems. To comprehend how an embryo develops, we have to decipher the language by which cells talk to one another. Moreover, we must decode the rules that dictate the social behavior of cells, to achieve the reproducible pattern of the embryo.

The Spemann and Mangold experiment triggered a burst of activity to identify the agents that instructed the host cells of the embryo to form a second head, including the brain cells. This led to a set of frustrating forays. What is the molecular nature of these instructions? Some researchers claimed that even boiled tissue could instruct the formation of head structures, indicating that the instructive signal is a heat-resistant substance. Others showed that simply separating between the cells could also give rise to the induction of brain-cell fates, bringing into question the actual need for induction and communication between cells. Thus, as revealing as Spemann and Mangold's experiment was, it was followed by more than fifty years of frustration, because researchers lacked the experimental means to unravel the cell language that ultimately gives rise to patterning.

Answers began to emerge when new tools of molecular biology were applied to developing frog embryos; this research helped uncover the mechanism of induction by the Spemann organizer. The general notion was that the organizer cells are different from their neighbors and express a special set of genes that can produce proteins that are secreted and can communicate with the neighboring cells. With the advent of molecular tools in the 1980s, it was possible to demonstrate that RNA molecules made by specific genes were exclusively present only in the region of the organizer. In molecular terms, it meant that there is a distinct entity within that region.

In the early 1990s, subsequent research uncovered that genes not only are expressed within that region but also encode proteins that are secreted outside the cell, and hence can indeed influence their neighbors. Could these be the molecules that mediate the communication envisaged by Spemann seventy years earlier? By transplanting these predicted molecules, scientists were able to mimic the results Spemann obtained by transplanting whole tissue. The results indicated that indeed these molecules represent the organizers in the striking response first observed by Spemann and Mangold.

What is the nature of these molecules? Because the field had seventy years to ponder this question, it is interesting to consider, in retrospect, the psychological aspects that influence the notions and predictions made by scientists. To reiterate, we are dealing with molecules that can lead to the generation of head structures, including the brain. We obviously think of neural functions and brain activity as the highest level of differentiation, even when they take place in a newt, because of the way we perceive the activity of our own brain. As such, it is intuitively obvious to us that forming a brain requires an intricate set of instructions.

When we consider the simplest scenario in which cell development takes place, the cells face two options. With only their original instructions they can progress on one track, guided by an "autopilot" mode along what we term a default path. Alternatively, if they receive new instructions from neighboring cells, they can alter their path and progress toward generating another cell type. So you can imagine scientists' surprise when they found out that the molecules that constitute the Spemann organizer are not inducers but rather inhibitors. In other words, these molecules are telling the nearby cells to be oblivious to other signals they encounter and as a "default" to embark on the process of forming a head. The cells that are not exposed to the inhibitory influence of the organizer instead receive and respond to signals from another set of molecules and are directed to make more mundane organs, such as the belly of the tadpole.

Can the cells of the newt embryo really form a head and brain simply by default? Clearly, this is not the case. One can imagine that an extremely elaborate set of instructions and decisions need to be followed to form such structures. So how can we reconcile this result with the inhibitory nature of the Spemann organizer? As I will elaborate later, the decisions that cells follow are executed in a successive manner. At each decision junction the choices are distinct and defined. It is the collective accumulation of decisions that eventually leads to the generation of so many distinct cell types and such elaborate structures. We can compare this to driving from one location to another. When you make the first turn, you are concerned primarily with the general direction of your destination rather than with the final street address.

Although the first decisions that cells make are crude, they are also the most important ones. The first decision the cells undertake in the newt embryo is whether to become the head or the belly structure. At this coarse level, the instruction to form a head is executed simply by inhibiting the alternative fate. Once a domain has been defined as the future head region, subsequent decisions will elaborate the pattern to generate a final structure. Indeed, highly orchestrated processes are required to determine and produce the intricate structures of the brain and eyes.

CHAPTER 4

How Cells Talk and Listen to Each Other

In the previous chapter, I discussed the implications of Spemann and Mangold's experiment of 1924, which revealed that cell fates are not predetermined during development. In the course of deciding what they will become when they "grow up," cells actually communicate constantly with one another. In other words, a cell is not oblivious to its environment as it carries out its destiny during development; it relies on the environment to receive continuous instructions. This feature resembles the contrast between the old class system, where a person's fate and occupation were determined from birth, and today's modern society in which we trust that a person's qualifications and interactions with his or her environment will shape the individual's ultimate destiny.

The implications of this notion for our perception of embryonic development cannot be overstated. To understand embryogenesis, we have to discover those instructions that cells are continuously transmitting to and receiving from one another. We must be able to comprehend the language of the cells. It means that the information in the DNA defines the final outcome only by setting the rules of interactions among cells, but there is no micromanagement.

By leaving the decisions open to the cells, the complicated process of development is further protected from fluctuations and perturbations. If each cell had a predetermined destiny, then a possible loss of any given cell could lead to dramatic patterning defects. In contrast, if cells have a wide range of fates they can assume, other cells can replace the loss of a given cell. This characteristic is similar to the declaration that there is no person who cannot be replaced.

As dramatic as Spemann and Mangold's experiment was (it is hard to be more dramatic than growing a second head on a newt embryo), its conclusions were left hanging for more than five decades. This gap occurred because, although the existence of cell communication whereby cells of one type instruct their neighbors (induction) was recognized, the agents that mediate this interaction remained unknown. The molecular components of cell communication could not be isolated biochemically. Despite the biological importance of these agents, many of them could be present in minute amounts, and the biological assays for their activity are not straightforward. In other words, without any clues as to their nature, it is impossible to fish out the specific components of cell communication from a mixture of all the elements of the cell. We can compare the situation to grinding up a cell phone in a blender. Without any previous knowledge, would you be able to identify the critical memory chip in the final mixture?

Using genetic tools, scientists eventually uncovered the pathways by which cells communicate. The genetic makeup of an organism is called the genotype, and it dictates the organism's final shape, or phenotype. Genetic analyses depend on the ability to generate and study mutations in genes. Because the sequence of DNA dictates the structure of the proteins that will form, it follows that a change in the sequence of a gene, arising from a mutation, should lead to a parallel modification in the structure of the protein. Depending on the nature of the change, this alteration can be subtle (like a single misspelled letter in a sentence) or drastic, leading to the formation of a nonfunctional protein. When mutations in the DNA modify a critical protein, an alteration in the appearance or functionality of the organism (leading to a new phenotype) is likely to follow. The classic example is the first mutation isolated by Thomas H. Morgan in 1910 in the fruit fly, *Drosophila*, in which an alteration in a single gene converted the color of the eye from red to white.

The isolation of this mutation, termed "*white*" after the phenotype it generates, exemplifies the logic of uncovering the genetic basic for biological processes. This mutation appeared randomly in a large population of flies and could be readily followed genetically over subsequent generations of

Figure 9. *Genetic mutations reveal the hidden function of genes*
When genes are removed, the defects that arise make the missing gene function apparent. Similarly, after a leaf imprints wet cement, its missing presence is recorded (B. Shilo, Hampi, India)

flies by virtue of the distinct eye color phenotype. To this day hundreds of research groups use this mutation in their research. The phenotype indicates that the gene plays a crucial role in eye pigmentation, since in its absence, no pigmentation is observed. The subsequent analysis of the *white* gene indeed uncovered the molecular basis for the production and transport of pigments to the eye. In other words, the phenotype arising from loss of a gene marks the gene as a central player in that process (fig. 9).

Consider, for example, the game we used to play as children in which one person initiates motions and the other people in the room imitate this person. If you step into the room, you do not know who initiated the motions you are observing. So, at random, you ask one person to leave the room. If the others continue their motions, obviously the person you selected was not the leader, and you go on to ask another person to leave the room. Only when the initiator leaves the room will the motions stop, and this will indicate that person's role.

But if we don't know beforehand which genes and proteins are relevant for the process under study, how do we generate mutations in these genes? The answer is that from the outset we don't target the mutations to distinct genes. Instead, we use the phenotypes that arise from the loss of or from defects in genes as the basis for identifying the genetic circuitry that dictates a particular biological process. The tissue that we study (for example, the coloration of the fruit fly eye or the shape of its wings) will define the genes that we will eventually select for further analysis. We can thus inactivate genes randomly in what we call a "genetic screen" and then follow up on the consequences by choosing the tissue or stage where we want to look for the arising phenotypes.

How is a genetic screen carried out? We can introduce mutations into the genome of an organism by randomly exposing it to ionizing radiation or chemicals that alter the fidelity of the DNA during replication. This "insult" to the genome will also alter the DNA that is carried by sperm and eggs. If we take the offspring of the mutated animals, we are stably transferring a given random mutation to all cells of its progeny, through the mutagenized DNA of sperm or egg. Such mutations can then be maintained in a stable manner, since they will always be transmitted to the next generation.

As you'll recall, the genome of multicellular organisms contains two complete copies of each chromosome, one derived from the sperm and the other from the egg. Thus, even if an essential gene is mutated and inactivated, in most cases its normal counterpart will remain active and "fill in" for the missing gene. In other words, cells and organisms can appear perfectly normal even when they carry deleterious mutations in one of their genes.

When a sufficiently large number of independent mutations are examined, the screen should identify most genes that are involved in the process at hand. We define such as screen as saturated. Let's return to the example I gave earlier of identifying the person who is the initiator of motions. Suppose there are twenty people in the room. If you ask one or two to leave, your chances of identifying the key person are slim. But if you ask all twenty people to leave, one at a time, you are bound to identify the initiator. You have "saturated" your screen because you tested all possible options.

It is important to highlight the open state-of-mind in which a researcher initiates such a genetic screen, especially when a complicated biological process is being examined. The process is too complex for us to have a precon- ceived notion that is likely to bear any resemblance to how the real mecha- nism actually operates. So the scientists carrying out a genetic screen should be unbiased and allow the mutations they isolate to reveal the story of how a given process is executed.

In the late 1970s, Christiane Nüsslein-Volhard and Eric Wieschaus carried out such a mutagenic screen using fruit flies. They generated a collection of about eighteen thousand fly strains, each harboring one or more mutations in its genome. We now know that flies have about thirteen thousand genes, so this collection most likely included representative mutations in most of these genes. Using this collection, they then carried out a genetic screen. The goal Nüsslein-Volhard and Wieschaus set for themselves was extremely am- bitious, because they wanted to identify the genes responsible for generating the embryonic pattern. Fruit flies represent a simpler model organism, and their process of embryonic development takes only twenty-four hours. Yet by the time embryogenesis is completed, the embryo contains fifty thousand cells, representing the full complement of differentiated tissues. In recogni- tion of their seminal findings, they were awarded the Nobel Prize in 1995.

Within the library of mutants, every strain carries one normal and one mutated copy of the gene that was randomly modified. But when males and females from the same line are mated with each other, some of their progeny will contain two copies of the defective gene. In cases where this gene is essential, no viable progeny will emerge from this combination. Genes can be indispensable because they mediate a host of processes, such as metabo- lism or structural roles within the cell. But in the context of the genetic screen for developmental mutants, the researchers' focus was on genes involved in the determination of embryonic patterns.

In normal fruit fly embryo development, the skin cells of the embryo are patterned so that they outline structurally distinct domains, or segments. Each of these segments displays unique structures, according to its position within the embryo, in both head-tail and belly-back axes. The skin cells se- crete a hard external protective layer, or cuticle, which we can think of as the outer skeleton of the embryo. The cuticle is composed predominantly of the nonliving substance chitin and is thus highly resilient and stable. Each cuticle's structure manifests the distinct properties of the cells that were responsible for its secretion. And it was these stable cuticle "shells" that helped crack the puzzle of embryonic development (fig. 10).

Even when mutant embryos die and disintegrate, their external cuticles remain stable and resilient. The researchers could collect the cuticles and analyze them to reveal the processes that went awry in the absence of a given gene, such as a missing tail or a duplicated head.

At the outset, it was not clear how many of the thirteen thousand genes in the genome of the fly are dedicated to determining the intricate embryonic pattern. The astonishing revelation was that mutations in only about two hundred genes gave rise to patterning defects. Since most of the genome had been screened, the obvious conclusion was that only a small number of genes is responsible for determining the complex structure and shape of the embryo. Furthermore, in many cases, mutations in different genes resulted in very similar defects, allowing this class of mutants to be grouped as players in the same process. In other words, the implication was that they act sequentially to carry out the same process.

Imagine the intricate process by which oranges are grown, harvested, and reach the supermarket shelf. Any defect along the way, from lack of adequate water and other growing conditions to mechanical problems in the truck transporting the oranges to the store will result in the same outcome: No oranges on the shelf! Each step thus defines an essential link in the same global process. Genes whose absence results in similar defects represent different steps in the same biological process. When these genes were subsequently isolated, the structure of the proteins they encode could be readily

Figure 10. *The genetics of the fruit fly uncovered the molecular language of cells* ▶
The outermost cell layer of the fruit fly embryo secretes a hard external cover or "shell" that is patterned according to the fate of the cells that secreted it. By screening eighteen thousand fly strains carrying random mutations in their genome, researchers identified mutations that give rise to patterning abnormalities. These mutations defined the genes encoding the proteins that mediate cell communication during development. Although the outer layer is only a "shadow" of the real embryo, it provided the clues for most of the components that comprise the blueprint for making the embryo.

Top: External "shell" of a fruit fly embryo (R. Schweitzer and B. Shilo, Weizmann Institute); *bottom:* Shadows can reveal the deeper underlying story (B. Shilo, Israel Museum, Jerusalem)

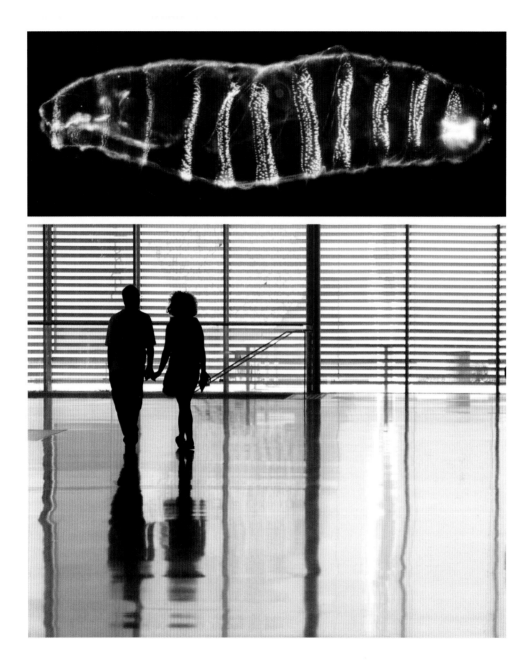

decoded. In many cases, these structures were extremely revealing, because they provided clues to the location of a given protein within the cell and how it might function at the molecular level. It indicated that these proteins may represent the molecular means by which cells communicate with one another. The mechanism Spemann postulated had now been validated and dissected. The "language" of cell communication was at hand.

How do we envisage this communication at the molecular level? Every cell is surrounded by a membrane, which is composed of multiple units of fatty acids, each consisting of a long chain of carbon atoms. The separation between the fatty acids and the aqueous environment of the cell defines the membrane as a distinct border between the cell and its surroundings. The membrane allows the cell to maintain a unique intracellular environment by selectively concentrating a host of substances within the cell and excreting others out of the cell.

Specific proteins are embedded within this membrane and fulfill a wide range of functions. The class of molecules relevant for our discussion is termed receptors. These receptors are responsible for the transmission of information across the cell membrane. The receptors are embedded within the membrane and protrude into the extracellular environment, where they contact hormone-like proteins that were secreted by other cells. Once contact between the receptor and the protein is established, the receptor's structure is altered. This happens not only in the external part where it associates with the hormone but also in the internal part that extends into the cell.

How is this change now transmitted all the way to the cell nucleus? Proteins can bind together, to alter their structure, stability, or location within the cell to form signaling pathways. In "relay race" fashion, these chains of interacting proteins transmit to the nucleus that the receptor was activated. In other words, a chain of interacting proteins transfers information from the membrane to the cytoplasm and then to the nucleus (fig. 11).

Although each signaling pathway has a similar goal, each uses a different set of proteins and distinct "tactics" to achieve it. In the signaling process many levels of modulation and regulation exist. Signals can be dampened as they are transmitted within the cell, or conversely they can be amplified when one protein activates many other proteins. One way of picturing the transmission and alteration of signals is Norman Rockwell's humorous and

Figure 11. The mechanism of cell communication
Cells receive signals from neighboring cells through their external environment. A
chain of proteins transmits the information through the cell membrane all the way
to the cell nucleus to induce expression of critical target genes. Such protein cascades
function as a "bucket brigade," here illustrated by a Playback improvisational theater
group, where actors transfer an imaginary object from one to another, modifying it in
the process (B. Shilo, Cambridge, Massachusetts)

folksy illustration *The Gossips*. Rockwell depicts a rumor being whispered and
transmitted from one person in town to the next. By the time it comes back
to the initiator of the rumor, she (in this case) can no longer recognize the
original message, which has been altered and modified along the way.

Having discovered signaling pathways through the genetic screens, the
researchers had two additional surprises in store. First, they learned that only
a handful of signaling pathways by which cells communicate with one another operate throughout the embryo's development. Five of them are very
prevalent and are used repeatedly at consecutive decision junctions; three
others are less prominent. Each pathway has distinct components and uses
different molecular modes. Yet they all relay information across the membrane and into the nucleus in order to alter the pattern of gene expression.

Second, they found that these pathways are highly conserved in evolution, most being shared among both simple and complex organisms. This means that cells in all multicellular organisms communicate using the same language and that this is also the language used by their common ancestors throughout evolution. The discovery of shared signaling pathways used by such a diverse array of organisms, from nematode worms and fruit flies to humans, changed the basic culture of scientific research. From segregated communities of scientists, each focusing on a different organism, it suddenly became apparent that everyone is actually studying the same molecular system. Processes as diverse as eye development in flies and human cancer are actually using exactly the same machinery. The universality of signaling fostered a broader, more global approach that scientists are now applying to biology.

How Do Simple Modules Lead to Complex Patterns?

The realization that the majority of cell communication takes place within a limited number of signaling pathways was rewarding. It provided a simple conceptual way to approach embryonic development and the intricate patterns involved in this process. Moreover, the universality of these pathways, as reflected in their conservation across diverse species, offered the exciting prospect of deciphering general developmental paradigms by examining a range of model organisms in parallel. In other words, cells in all multicellular organisms are speaking the same language.

How does communication among cells actually lead to the generation of patterns? In essence, cells communicate and coordinate, using rules that are encoded in their genome. Let's start with a simple example, in which the use of a single communication pathway allows cells to generate a repeated pattern in a highly reproducible manner.

One of the major communication pathways is termed "Notch" because when the activity of this pathway is compromised, notches form in the wings of adult fruit flies. The protein that triggers the pathway is anchored to the membrane of the cell that displays it, so that it can only mediate communication among cells that are physically touching one another. This pathway functions in multiple biological settings. In all of them, this protein activates a membrane receptor in the adjacent cells and gives them the following instruction: "I am the cell that will form a particular cell type (for example, a neuron, which is a specialized cell that sends out and receives electrical sig-

nals), and you will not." This simple communication will eventually generate regular spacing between the neurons (fig. 12).

Imagine a layer of cells that are identical not only in their genetic makeup but also in their state and status. Although they are equivalent, there are always random differences among cells in the amount of the proteins they express; *the level* of the protein activating the Notch pathway is no exception. For a given cell, once this protein level randomly rises above a critical value, this cell will become a neuron. In parallel, this neuron instructs its immediate neighbors not to become neurons. Over time, the layer of originally identical and nondifferentiated cells will form a grid of neural cells, which are separated by non-neural cells at regular intervals. A pattern was formed using a single communication pathway and a simple set of rules (fig. 13).

What are the consequences of malfunctions in such a pathway? If the cells are missing the receptor, they become "deaf" to the instructions transmitted by their neighbors. In this case, all cells will form neurons, because they all have the potential to do so, and are oblivious to the signals from the adjacent cells telling them not to become neurons (fig. 14). This dramatic consequence highlights the essential role of this communication pathway in dictating the final pattern.

The enormous simplification and universal use of communication channels among cells come at a price. The problem at hand is now actually reversed. We know that embryos are extremely complex and can contain thousands of different cell types. How can this enormous complexity be dictated by such a simple and restricted set of communication channels? What are the

◀ *Figure 12. How cell signaling establishes a repeated pattern*
The compound eye of the fruit fly is made up of eight hundred repeated units called ommatidia. During eye development, patterning takes place progressively from the bottom to the top of the picture, generating a new row of ommatidia every two hours. Once an ommatidium is established, it signals to inhibit cells at the immediate radius from assuming a similar fate. The newly formed ommatidia become spaced at fixed distances, just outside this radius. Once created, they will now generate their own inhibitory spheres, to outline spacing of the next row of ommatidia that will form.

Top: Establishment of the founder cells (blue) for ommatidia during eye development (A. Shwartz and B. Shilo, Weizmann Institute); *bottom:* Pigeons sitting on a railing establish a fixed spacing (B. Shilo, Moss Landing, California)

Figure 13. How cell signaling establishes a repeated pattern ▲
The outermost cell layer of the fruit fly pupa displays a regular arrangement of nerve (red) and epidermal (black, outlined in white) cells. All cells are initially equivalent. Yet random fluctuations in the level at which each cell produces a signal allow a given cell to accumulate sufficiently high levels that make it a nerve cell. In parallel, it instructs its immediate neighbors to become skin cells. Eventually, an ordered array of nerve cells interspersed by skin cells is generated.

Top: External cell layer of the fruit fly pupa with regularly spaced nerve cells (red) (S. Yamamoto and H. Bellen, Baylor College of Medicine and Howard Hughes Medical Institute); *bottom:* Sunshades distributed at regular intervals on a beach on the Greek island of Lefkada (B. Shilo)

Figure 14. When cell communication breaks down ▶
When communication among the cells is disrupted through a manipulation that genetically inactivates the receptor, the cells in the patch at the center become "deaf" to the signal from their neighbors. Since these cells do not perceive the inhibitory cues, they all become nerve cells.

Top: External cell layer of the fruit fly pupa with regularly spaced nerve cells (red) and accumulation of nerve cells where the receptor is missing (S. Yamamoto and H. Bellen, Baylor College of Medicine and Howard Hughes Medical Institute); *bottom:* When sunshades are placed without regard to their spacing with adjacent sunshades, a continuous and disorganized "mutant" clustering is generated (B. Shilo, Tel Aviv)

design principles of the process that generates such a reproducible pattern using only basic pathways?

How Do Sequential Decisions Refine and Elaborate Patterns?

Because cell interactions define the pattern, its complexity is most likely generated by elaboration of those interactions. To avoid errors and allow cells to make the correct decisions, simplicity is an essential requirement at each junction (fig. 15). Imagine that you are traveling from one place to another in an unfamiliar city without a map and before GPS (global positioning system). You would probably ask the first person you saw for instructions, and he or she would point you in the general direction to the next signpost, where you would then ask the next individual. Having reached this second point, you would be directed further to the next junction, and so on until you arrive at

Figure 15. *Don't make all the decisions at once*
When one reaches an intersection where decisions have to be made and choices taken, it is hard to decide which way to go when many alternatives are presented at once. Cells in the embryo are faced with only a small number of alternatives at each junction (B. Shilo, Davis Square, Somerville, Massachusetts)

your destination. Your journey started from a general instruction regarding the overall direction and ended at a specific location.

Several aspects in this analogy are relevant for how developmental complexity is generated. The first point concerns the nature of the instructions you obtained along each step of the way. When you are starting out and are still far from your target, the instructions can be fairly coarse, pointing you only in the general direction. As you approach your destination, however, the instructions need to become more specific, relating to the precise street and eventually the specific house number.

The second issue concerns the significance of the instructions based on when you received them. The initial instructions, although coarse and general, were the most critical in the sense that, if you had been pointed in the opposite direction, you would have proceeded on a dramatically different course. Therefore, as you advanced, you received more specific instructions, but the first ones remained crucial in determining the final global outcome.

The importance of the initial direction can also be highlighted by the analogy to career development, where an individual pursues education and training for a future occupation and job. First, the person chooses only a general area of focus. Nevertheless, that decision is perhaps the most crucial one in terms of defining the milieu in which the specific occupation will eventually be chosen. Many people identify a charismatic high school teacher they were fortunate to have as the trigger for their current career.

How do cells make successive choices? The initial set of instructions is crude but critical. For each axis within the embryo, there is an event defined as "symmetry breaking," in which a symmetrical form is converted to a polarized one. The symmetry-breaking event could be random, meaning that it does not depend on previous positional information. For example, the fertilization of the frog egg takes place in water, and the sperm can penetrate at any position along the top of the egg. The random location where it penetrates, however, will break the symmetry within the fertilized egg, so that it will define the future head region of the embryo in the region opposite to the site of sperm entry.

Once the original symmetry has been broken, how is pattern elaborated? Creating the initial asymmetry gives rise to a change in the distribution of components within the cell. For example, the cell nucleus might be moved to one corner, the orientation of intracellular "cables" within the cell may be altered, or particular proteins could be concentrated in one part of the cell. Typically, the symmetry-breaking event takes place when the embryo is still a single cell. For example, this event might occur at the fertilization stage or during the formation of the egg even before it is fertilized, because various cellular components within the egg are positioned asymmetrically. A solitary symmetry-breaking event taking place within a single cell will define the future axis of the entire embryo. This singularity in an event that may depend on a random process is important because it eliminates the need for elaborate coordination at a stage that is so critical for the entire course of embryonic development.

Once the initial asymmetry is generated, it is stabilized and becomes the basis for the elaborate patterning of the embryo. In terms of the early embryo, this first asymmetry can be manifested in a variety of mechanisms. In one case, it can lead to a local concentration of RNA molecules that encode critical signaling components. The result is a localized translation of these RNAs and the subsequent local activation of a particular pathway. In other cases, the asymmetry may be manifested in the local activation of proteins that modify other proteins essential in a signaling pathway. Again, this manifestation would lead to the confined activation of that specific signaling pathway. In still other cases, the RNA being localized encodes a protein that triggers gene expression, thus leading to a local activation of target genes.

Although the symmetry-breaking event and the local processes it triggers may be highly diverse among organisms, the final general outcome is common. By biasing the local activation of a critical patterning pathway, the outcome will eventually be transmitted by this pathway to the nucleus. This transmission will culminate in the expression of a given set of genes in that part of the embryo and not in other domains. A transient asymmetry has now been converted to a stable change that is maintained by the embryo.

This initial asymmetry has a global effect on the entire embryo. It determines either the position of the head versus the tail or the belly versus the back. After it is stabilized by local protein synthesis within the embryo, different regions will begin to express distinct sets of proteins. The first response to the broad asymmetry is to subdivide the embryo into a handful of distinct zones, each expressing a different set of genes. In subsequent steps, each zone will be further patterned and divided into smaller distinct domains. The initial steps impinge on the whole embryo, whereas subsequent stages are confined to smaller domains, with each stage providing more elaborate patterns within its limited zone. This temporal order of patterning generates a clear hierarchy between the genes, where the genes that regulate the earlier steps have the most global and prominent effect on the resulting embryo shape.

Consider, for example, how patterning of the head-to-tail axis takes place in the fruit fly embryo. When each egg develops within the female, it is positioned at one end of the developing egg chamber, while fifteen "nurse" cells, which nourish the egg as it grows in size, are placed at the other end. Because the embryo will develop outside the mother, the egg contains all the food required for the first twenty-four hours after fertilization, as well as the critical patterning cues. In this case, the symmetry-breaking event is the positioning of the egg at one end of the chamber, because all the substances it receives from the nurse cells flow into the egg in one direction. One of the RNAs the nurse cells provide to the egg is called *bicoid*. This particular RNA is trapped and actively positioned at the site where it enters the egg.

The fruit fly embryo has a novel feature in which membranes separating the nuclei are formed only after embryogenesis has been going on for at least two hours, at which time the embryo already contains five thousand nuclei. Thus, in spite of the embryo's having many nuclei, there are no partitions between them, and the whole embryo can be regarded as a single, giant cell measuring 0.5 mm long and 0.2 mm wide. This cell is an object you can actually easily see with the naked eye, hold with tweezers, and manipulate.

Once the fruit fly egg is fertilized, the RNA molecules it contains will begin to produce protein. Because the *bicoid* RNA accumulated at one pole, which represents the future head of the embryo, it will make many copies of the protein only in that region. In the absence of barriers among the nuclei, the *Bicoid* protein molecules now diffuse from the original site of formation into other parts of the embryo. Although biologists and physicists are still debating the precise mechanism of diffusion and its speed, the result is that a continuous gradient of the Bicoid protein is generated. This means that there is more Bicoid protein closer to its site of production and less protein the further from the source, toward the tail of the embryo. Every region along the head-tail axis of the embryo is now roughly defined by the concentration of Bicoid protein it displays.

At this point the embryo stops relying on the RNA and protein molecules that were deposited into the egg and starts to make its own materials. By now the embryo contains several thousands of nuclei that are evenly distributed. Bicoid protein enters each of these nuclei and controls the expression of genes. (See Chapters 2 and 7, as well as later in this chapter, for more detail on how such regulation takes place.) Here it is enough to say that different levels of Bicoid protein can direct the expression of distinct genes. Thus, by virtue of the spatial distribution of a single molecule, Bicoid protein, the embryo is now subdivided into several distinct zones, each expressing different sets of genes. These genes, termed gap genes, also encode proteins that are targeted for the nucleus to direct gene expression. This transition represents the first step in patterning.

In the next stage, the borders of expression of adjacent gap genes will be sharpened, such that each cell will express only one type of gap genes. In addition, the positional information provided at the borders between different gap genes becomes the basis for expression of the next cohort of genes. Thus, in two steps, and with relatively simple rules, there has been a progression from a coarse gradient of the Bicoid protein to defined zones within the embryo that now express specific genes at their borders. This pattern will eventually define the segmented pattern of the embryo (fig. 16).

Figure 16. *Accumulation of simple decisions* ▶
If embryonic fates are dictated by only a handful of signaling pathways that are used reiteratively, how is the enormous complexity of the organism generated? Although the number of alternatives is limited at each decision junction, progressing from one choice to the next generates complexity and gradually refines the pattern.

Top: Patterning for the head-to-tail axis of the fruit fly embryo. Initially a protein (left panel, blue) is distributed in a coarse gradient, with the highest levels in the future head (top). According to the concentration of this protein in different parts of the embryo, distinct genes are expressed in each region, such as the gene in green that is expressed in two defined zones. The interaction and small overlap between proteins that are expressed in zones will now induce expression of a gene (red) in a repeated pattern of seven stripes (J. Reinitz, University of Chicago); bottom: As the game of dominoes progresses, each new cube will increase the complexity of the cube assembly that is generated (B. Shilo, Old City of Jerusalem)

How Do Signaling Pathways Cooperate to Relay Information?

Identifying the handful of signaling pathways that mediate all communication events among cells during development was vital in showing researchers the language of cells. With the genes and mutants for these pathways at hand, scientists could now address the different junctions in which they are used during embryonic development. Each pathway guides tens, if not hundreds, of developmental decisions made by the cells during embryo formation.

Each of the signaling pathways relays information from the extracellular environment into the cell by crossing the cell membrane and eventually transmitting this information into the nucleus to direct a new program of gene expression. This fixed relay system begs a new question. If a given pathway participates in numerous decisions and guides multiple differentiation choices, it cannot carry the full set of instructions on its own. How are simple pathways used to transmit and direct such a wide array of choices? This is the time to consider how gene expression is activated at the nucleus.

The information that a gene contains to produce protein represents only part of the gene. The other adjacent DNA segment, which is typically even longer than the former, is involved in regulating the gene's expression. To produce RNA from the DNA template, an elaborate machinery of proteins is required. This large protein complex, called RNA polymerase, associates with the DNA, slides along it, opens up the double-stranded helical structure, and introduces the building blocks of RNA based on the DNA template. It can produce per minute an RNA chain of about three thousand units. When RNA is being actively produced, multiple RNA polymerase complexes can be "running along" the same gene, one on the heels of the other. Once the complex binds the DNA at a given gene, the process of making the RNA usually follows immediately. It is the actual binding of the complex to the DNA that needs to be regulated.

This is where the regulation of gene expression becomes relevant. The RNA polymerase complex has the potential to copy any one of the twenty-five thousand genes in the human genome. But it operates only on the genes that are marked by virtue of specific proteins that associate with the regulatory portion of these genes. These proteins provide a structure for the RNA polymerase complex that is functionally equivalent to a landing pad. So the question of controlling the production of RNA boils down to the determination of specific landing pads on the DNA that would be distinct for each cell type and every developmental junction.

The basic elements of these landing pads are proteins, each binding a specific short stretch on the DNA, composed typically of six to ten nucleotides. As such, they are extremely flexible and can be compared to beads on a string. Like a string, the DNA structure is highly bendable. In fact, the folding of the DNA brings together the proteins that are bound to the DNA at distant sites. Even proteins associated at sites that are as distant from each other as thousands or tens of thousands of bases on the DNA can be positioned next to each other by this folding. When consolidated, these proteins form a landing pad for the RNA polymerase.

The arrangement of proteins on the landing pad is highly versatile, and many combinations of proteins can each generate an effective site for the binding of RNA polymerase. This modularity provides an effective way to integrate information and signals. As mentioned, each of those few developmental signaling pathways culminates in a protein that is bound to specific sites on the DNA. When the landing pads contain a combination of binding sites triggered by different pathways, they effectively integrate information. This process is analogous to classic thriller movie scenes in which two people have to insert their keys simultaneously to arm an atomic bomb (that will, we hope, be disarmed later on by the superhero).

The integration of signaling pathways at the level of the genes they activate has several important consequences. First, it is an effective way to generate diversity. The human genome contains roughly twenty-five thousand genes. If creating a landing pad for a given gene involves binding the proteins triggered by two of these pathways, then only when the two pathways are simultaneously activated will the gene be expressed. A different blend of genes that will be triggered can be defined by combining another pair of signaling pathways, and so on. As one can imagine, there will be multiple possible combinations among five pathways, in pairs, triplets, or more, thus generating numerous distinctions among genes that are recognized by given pathway combinations. For example, in winter, we can distinguish people in a crowd who have both red hats and black gloves, and in another instance

those who have red hats and blue jackets. In each case a different group of people will be selected, with some possible overlaps.

The second consequence of combining the input from several pathways relates to the finer spatial and temporal resolution of gene expression. Suppose that each of the pathways is activated in a different domain. The target genes will be expressed only in the cells where the two domains spatially intersect. Similarly, if the pathways are activated at different times, with a narrow window of overlap, the target genes will be expressed only within this distinct time window. The ability to combine the input from several pathways allows the definition of very distinct groups of genes that will be induced at each setting and can confine the expression of these genes to a narrow spatial and temporal window (fig. 17).

Can the Same Signal Be Interpreted in Different Ways?

Landing pads, however, are often functional only in the presence of additional proteins that define the tissue context. When a cell is already differentiated, how does it "know" in a global sense what type of cell it has become? I mentioned earlier that the identity of a cell is reflected in the set of genes it will express. We thus have to consider the assimilation of the new identity, reflected in the genes that a differentiated cell will now choose to express, in the context of genes and proteins that this cell is already expressing.

Certain genes, known as master regulatory genes, define and maintain tissue identity. In other words, the proteins these genes encode will trigger the entire cascade of genes subsequently needed to execute the identity of

◀ *Figure 17. Combinatorial interactions among signaling pathways*
Although only a handful of cell communication pathways guide development, two or more can converge to regulate the expression of a given target gene. This intersection between pathways defines more precisely the time and location at which a target gene will be expressed.

Top: Two signaling pathways in the fruit fly embryo (red and blue) induce the expression of specific target genes at the points of intersection between them (E. Furlong, European Molecular Biology Laboratory); *bottom:* Similarly, in a painter's palette, the basic colors can be mixed to generate a dazzling array of combinations (B. Shilo)

that tissue. It follows that, if such genes are experimentally expressed in another place, they will have the capacity to produce a whole organ in this new position. A dramatic example of such a master control gene is *eyeless*, which determines the position of the fruit fly eye.

The name can be confusing, so this is a good place to explain how genes get their names. When the gene that gives rise to the eye is defective because of a mutation, the resulting outcome is a fly without eyes. Because the gene was first identified by virtue of its mutant phenotype, it was given the name *eyeless*. However, the normal role of *eyeless* is exactly the opposite: it defines the position of the future eyes. The structure and function of the Eyeless protein is amazingly conserved in the evolution of multicellular organisms. In humans, a gene highly similar to *eyeless* (termed *congenital aniridia*) dictates eye development, such that in its absence the iris of the eye is reduced or missing.

So what happens if the Eyeless protein is experimentally manipulated, such that it is expressed in the future wings and feet of developing fly larvae? Amazingly, all of these organs grow eyes. The new eyes cannot see because they are not wired to the brain. But at the level of each unit, they look like normal eyes. This result relates to the observation I made earlier in describing the Spemann and Mangold transplantation experiment in the newt. The initial manipulation, whether the transplantation of the future head region or the expression of Eyeless in the wrong locations, is crude. Yet it mimics the normal course such that these master regulators now direct the succession of events according to the rules of the original organ. Hence, the final structures that are formed are accurate and indistinguishable from the natural tissue.

How does a tissue master regulator affect how the cells within that tissue perceive their identity? The master regulator is typically a protein that regulates gene expression by binding the regulatory DNA of genes and contributing to the generation of the landing pads for the machinery that produces RNA. The developmental signaling pathways are activated numerous times and in a variety of tissues and cell types. The genes they trigger usually require that the landing pad provides not only the signals from the developmental pathways but also a mark of the tissue identity, as a crucial element in creating a functional landing pad. It follows that, within a given tissue, the developmental pathways will only be able to trigger genes that are also marked by the tissue master regulator. Thus, even though the same developmental pathways may be used in all tissues, they will induce different genes in each tissue, depending on the markings of the master regulators.

This feature contributes significantly to the versatile use of the restricted number of signaling pathways. We can compare it to an electric current to which a variety of devices are connected. The resulting action will be totally different in the case of a toaster, a television, or a lightbulb, since the context of the device is what determines the outcome. Similarly, in different tissues, the signaling pathways can be regarded as switches, whereas specific tissue regulators determine the actual context of their activation. In other words, a cell's history will determine its response to a given signaling pathway. This history is reflected in the repertoire of tissue master regulators it contains and will determine the context of the response. Because cells change their properties over time, how they perceive the same communication channel may differ according to the new identity they have acquired (fig. 18).

What Is the Significance of Variations on a Repeated Module?

Although animals vary greatly in outward appearance, even the casual observer can often identify repetitive structures that provide some law and order. In vertebrate organisms, for example, the vertebrae are repeating units. By contrast, arthropods, which include the crustaceans, insects, and arachnids, are made up of repeated segments. In the case of organisms such as the

Figure 18. The history of a cell affects its responses ▶
As cells undergo successive cell fate decisions, their internal composition changes. Based on the specific array of proteins they express, the cells will respond in a distinct way to the same repeated signal. Thus, cell history and context generate a diversity of responses.

Top: Cells of a quail embryo were grafted to a duck embryo (generating a "quck"). The left-hand panel shows the quail cells in black. The right-hand panel illustrates that they are expressing a gene (white) that the neighboring duck cells are not expressing. Thus, although all cells sense the same environment, the quail cells' history leads them to express distinct genes (R. Schneider, University of California, San Francisco); *bottom:* A barbershop in Somerville, Massachusetts, owned by Central American immigrants. The background and history of these immigrants shape the way they perceive the local environment (B. Shilo)

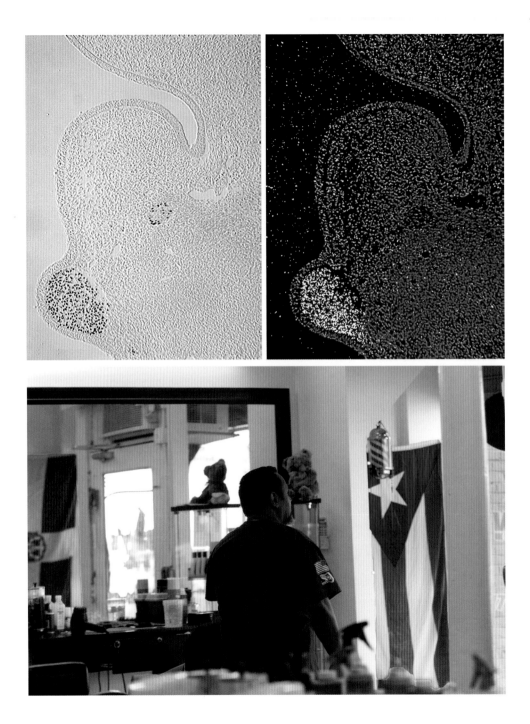

Figure 19. Variations on a repeated modular unit
Distinct structures can be generated along the body without the need to invent each one from scratch. Both vertebrates and invertebrates form a body that is initially divided into repeated identical units. Variations in gene expression in each of the segments, according to its position along the body axis, will subsequently create a diversity of structures.

Top: Killer whale skeleton at Northwest Building, Harvard University (B. Shilo); *bottom:* Although all these hotel porches have the same furniture, variations in their arrangement generate diversity in appearance (B. Shilo, Tel Aviv)

Figure 20. *More variations on a repeated unit*
Top: Fruit fly larva showing a subtle variation in the repeated structure of nerves (green), connecting to the muscles (blue) in each segment (R. Sowade, B. Göllner, H. Aberle, and C. Klämbt, Münster University); *bottom:* Women weeding a field. Although all the women appear similar in custom and activity, variations in dress color and body pose are apparent (B. Shilo, Rajasthan, India)

centipede, this repetition is more obvious because the parts are similar to one another. For other insects such as flies, the differences in the morphological appearance of segments may mask the basic design derived from a common repeated structure (figs. 19 and 20).

How does the concept of a repeating module relate to the process of patterning the organism? It means that diversity among organisms can be generated by modulating two critical features: the number and the type of the repeated unit. First, consider variations in vertebrae number. Although vertebrae are the building blocks of all animals with backbones, there is an enormous difference in the number of vertebrae present in different species. Frogs have approximately ten, humans have thirty-three, and snakes have more than three hundred. So even as the same mechanism of subdividing the body into segments is maintained, alterations in the quantitative parameters of the process can result in very different numbers of segments.

The generation of a repeated unit in a developing vertebrate embryo has puzzled researchers for decades. During embryogenesis, the embryo of the frog, human, or snake continuously grows in size by adding cells to the tall part of its trunk. In a process of continuous uniform growth, how can discrete units that will subsequently develop into vertebrae be defined? The answer is found in a biological clock mechanism that operates, in a way, like a regular clock that divides the continuous time dimension into discrete units. The clock has two associated parameters—cycle duration and the number of cycles that are employed. Although the molecular details and the origin of these oscillations are not totally understood, the conceptual basis for the operation of a biological clock is that one gene (A) is expressed and the protein it encodes accumulates. This protein in turn induces the expression of another gene (B). But the B protein represses the expression of gene A. This process leads to cycles in which A is expressed, leading to accumulation of B, and subsequently to the shutdown of A. Once protein A levels are low, the levels of B also recede, and as a result protein A will now go back up, and vice versa. The time it takes each of the proteins to accumulate and disappear in the cell will determine the total duration of the cycle.

The nonsegmented region of the embryo grows continuously. At each stage, a small domain within this region is specified to be competent to produce the next segment by integration of converging signals. A new segment will form every time the biological clock within this domain goes through a cycle. The number of vertebrae will be determined according to the number of such cycles during embryogenesis. The main point is that by maintaining the same mechanism but modulating only one parameter, dramatic changes in pattern can take place. For example, by extending the time window over which segments can be formed or by speeding up the pace of the biological clock within the same time window, more cycles and hence more vertebrae will be formed. This is a point we will come back to in a different context— namely, that dramatic pattern alterations can take place without the need to invent a completely different patterning mechanism.

Having formed a given number of repeated structures that are typical to each species, whether vertebrae or arthropod segments, how is each of these units endowed with a distinct pattern? Why does the thoracic segment, which gives rise to wings and legs in the fly, differ from an abdominal segment? William Bateson, who published *Materials for the Study of Variation* in

1894, tackled this issue. Bateson studied specimens from natural collections and anatomy departments that showed abnormal patterns. These specimens included flies and bees with legs where there should have been antennae, frogs with altered number and types of vertebrae, and so forth. He distinctly cataloged the specimens into ones in which the numbers of repeated structures were abnormal and ones in which the fate of a repeated structure was converted to another. He called the change in the structure of the body part homeotic (from the Greek word *homeo*, meaning same). Bateson was interested in an organism's ability to obtain large jumps in the evolution of patterns. Although such alterations are not likely to give the organism an evolutionary advantage, the fact that these large jumps do occur points to an underlying molecular control mechanism dictated by specific genes.

The molecular insights into the identity of the repeated units emerged again from studies of fruit flies. Different "monsters" were generated by a mutation in a single gene. In one case, Nobel Prize laureate Edward B. Lewis (who shared the 1995 Prize in Physiology or Medicine with Christiane Nüsslein-Volhard and Eric Wieschaus) generated a four-winged fly, in which the third thoracic segment, which normally does not make wings, develops just like the segment ahead of it. In another case, flies that grew legs on their head instead of antennae were observed. Despite the dramatic alterations in pattern, these aberrations were each the result of a single mutation. It was thus clear that the mutated genes play a central role in a switch mechanism that ultimately determines the shape of a whole segment, with all the appendages it produces. They are master regulators in the sense that they harness tens if not hundreds of genes that ultimately shape the segment.

Analysis of the genetic basis for these homeotic pattern alterations in the fruit fly eventually provided much more than originally expected. In fact, the study of these genes has unraveled the universal code for the patterning of all multicellular organisms. When the first fruit fly gene encoding a homeotic patterning gene was isolated, a portion of it displayed a sequence of DNA, termed the homeobox, which encodes a domain of sixty amino acids within the full, much longer, protein. The sequence and structural features of this domain bear the hallmarks of proteins that interact with DNA, indicating that the protein is involved in binding DNA and creating the landing pads for the expression of target genes. This structural feature is in line with the capacity of a single homeobox-containing protein (or Hox for short), to dictate segment identity by driving the expression of a large battery of genes. In other words, a Hox protein creates landing pads that end up setting the stage for genes that will be expressed in a particular body part.

After identifying the homeobox as the functional unit (or "business end") of the first identified Hox protein, scientists went on to investigate whether other homeotic genes that specify different parts of the body have common structural features. It was indeed established that the other homeotic genes also contain sequences that are similar to the original homeobox. This feature resembles the placement of the same designer's logo on a variety of fashion objects, including clothes, perfumes, and watches. The logo indicates that, despite external differences, these items have something in common, indicating that they originate from the same source. Similarly, identification of the homeobox in different homeotic genes indicated that all of them originated from a duplication of an ancestral Hox gene. Furthermore, identification of the homeobox as the common region among several homeotic genes provided a powerful molecular handle or tag to isolate additional Hox genes from the same organism. We can compare this, if you will, to pulling needles out of a haystack by using a powerful magnet. Having identified a specific and unifying feature for the needles facilitates their isolation from the surrounding haystack.

It was rewarding for scientists to find out that each Hox gene is indeed expressed within the region where pattern abnormalities are observed in its absence. Essentially, when a Hox gene is missing, the relevant segment assumes the pattern of the segment preceding it. A famous example is Lewis's creation of a four-winged fly. Normally, flies have two wings produced by the second thoracic segment. The following segment produces small stubs called halters. When the Hox gene specifying the third thoracic segment (termed *ultrabithorax*) is absent, this segment assumes the fate of the second segment, making another pair of wings.

But the most striking finding, which the genetic mapping studies had already indicated, is that the Hox genes are arranged within a contiguous block on the chromosome and that each Hox gene's position within the cluster corresponds to the relative position of the body part it dictates. Thus, the Hox gene specifying the head is located before the Hox gene specifying the wing.

This is a very unusual organization. Typically, the regulation of genes relies primarily on regulatory sequences that are in their immediate vicinity. Genes encoding proteins that participate in the same process will most likely be located on very different chromosomal sites. This is similar to a multi-volume encyclopedia, in which, for example, the entries for Paris and France are in different books. The unique organization of Hox genes suggests a co-ordinated regulation of their expression that is linked to their physical location on the chromosome, such that genes that are located at one end will produce RNA before the genes that lie after them.

The surprises did not end with the fruit fly. Using the common module of the fly Hox genes, researchers could then isolate Hox genes from an ex-tremely wide range of other multicellular organisms, from the lowly hydra to humans. Technically, this was achieved in a way similar to the method I had used three years earlier to isolate from the fruit fly genome genes that are similar to human cancer-causing genes. The sequence of one homeobox is radioactively labeled and placed on a collection of the entire genome. It will stick only to sequences that are similar to it, thus allowing their identifi-cation and subsequent isolation. Again, this resembles employing a powerful magnet to pull needles out of a haystack, by virtue of a common feature that specifically marks them.

As in flies, in the other organisms Hox genes are also clustered, and their position in the cluster corresponds to the localization of the body part they determine. Thus, the Hox module enables us to understand the universal way in which the basic body plan is specified in very diverse organisms. Each Hox gene induces tens and even hundreds of genes that eventually dictate the spe-cific identity of the module.

CHAPTER 6

How Can a Single Substance Generate Multiple Responses?

Exerting a large-scale effect is central to patterning, especially at the early stages of embryogenesis. Patterns are generated and subsequently elaborated in a gradual and hierarchical manner. A global coarse asymmetry is first provided to the embryo or the tissue and then refined by pathways that respond to it. In the previous chapter I described the Bicoid protein, which is distributed in a graded manner within the early fruit fly embryo, dictating the future global body plan along the head-to-tail axis. The concentration of this single protein at each region will determine the future form of that domain.

Cell communication is the central mechanism behind the generation of patterns during embryogenesis. It is easier for us to imagine cells communicating at short range, where we have one cell sending a signal and the neighboring cell receiving the message. This is indeed the way signaling operates in the context of a pathway I described in Chapter 5, the Notch pathway, where the signal is actually attached to the membrane of the sending cell and hence cannot travel to more distant cells. Although close-range signaling strategies are important, they are limited in their capacity to drive global changes.

I now focus on the mechanism of the initial global signal. How do cells know their position within a developing embryo or organ and choose their specific developmental path? As a metaphor we can use the experience of a tourist in a big city, who may not know precisely where he or she may be at a given time but can estimate readily their distance from a prominent, visible landmark — for example, the Eiffel Tower or the Empire State Building.

Morphogens, or shape producers, are the substances that mediate the regulation of large domains during development of an organism. A morphogen can elicit several distinct responses, such that, with one signal, an entire field of cells will be subdivided into numerous distinct domains, each expressing a characteristic assembly of genes. (Alan Turing, who was responsible for breaking the German codes during World War II and a pioneer in computer science who after the war turned his interests to biology, coined this word.) A morphogen is a diffusible substance produced by a single cell or a limited group of cells. Once the cell secretes the morphogen, the substance is distributed in the extracellular space or along the outer membranes of the cells. The levels of morphogen decline as they get farther from the source, because morphogen molecules are diluted by diffusion and are also actively removed by degradation or uptake into the cells encountered along the way. As a result, local production of the morphogen will generate a graded distribution across the neighboring cells, at distances up to several tens of cells away from the source. This phenomenon represents an effective way to distribute positional cues in a global manner (fig. 21).

This process is conceptually similar to the distribution of the Bicoid protein, although Bicoid is distributed within the early fly embryo, which represents one giant cell, and morphogens are typically dispersed in the space outside the cells. A number of secreted molecules can function as a morphogen, each activating a different signaling pathway and in a different tissue or context.

Once the morphogen is distributed along the external surface of the cells, how is the information it carries conveyed to the cells? The cells need to perceive not only that they are encountering the morphogen but also *how much* morphogen they sense. In other words, detecting the morphogen does not trigger an on-off switch but can actually give rise to a variety of responses based on the level that is being sensed. This process is no trivial feat, because each cell has to assess how much morphogen it encounters without using its neighbors as reference points for cells that have received more or less signal.

As a metaphor for sensing a morphogen gradient, consider walking in Manhattan in the vicinity of a Macy's Department Store branch. When we are a few blocks from the store, we may occasionally encounter someone holding a Macy's shopping bag. But as we get closer to the "source," the number

Figure 21. *Transmitting positional information to cells*

How can an entire layer of cells be patterned at once by only one signal? Signaling molecules termed morphogens come into play here. These proteins are secreted by a restricted number of cells and spread to the surrounding space, up to tens of cell diameters away. The morphogen will have a higher concentration closer to its source. Cells can sense the level of morphogen they are exposed to and therefore where they are in the field with respect to the source. Cells will differentiate into distinct fates depending on their position within the field.

Top: Morphogen gradient of the Dpp protein (white) in the cells of the fruit fly larva that will give rise to the adult wing. The morphogen is produced and secreted on the left side (O. Wartlick and M. Gonzales-Gaitan, University of Geneva); *bottom:* A museum guide's voice provides a source of information and spreads over a distance. Students closest to the guide will hear her the loudest, whereas students in the back rows will hear the guide at a lower volume (B. Shilo, Mumok, Vienna)

of people carrying such bags increases, telling us we are heading toward the store.

What Keeps Organs in Proportion?

Signals at different distances from the morphogen source will give rise to diverse tissues, depending on their position. In the case of the transplantation experiment into newt embryos that Hilde Mangold and Hans Spemann analyzed (Chapter 3), the highest morphogen levels gave rise to the nervous system and the brain, whereas lower levels gave rise to skin and internal tissues. But now there is another complication! The overall size of the field being patterned can vary. For example, if female newts are exposed to better nutritional conditions, the eggs they lay, and consequently their embryos, are larger. Conversely, under poor conditions, the newts lay smaller eggs. So embryo sizes will vary. This variation is not detrimental as long as the embryos are functional.

To achieve reliable and functional patterning, the relative proportions of the different body parts must be preserved. If an embryo is smaller, the limbs must also become proportionally smaller. Suppose that a given morphogen value of, say 50 percent, specifies the position of the arm along the body length. This value should be at different absolute positions in the small and large embryos. Maintaining the relative proportion of embryos and organs, which adjusts to changes in overall size, is defined as scaling. The original Spemann and Mangold experiment, in which two well-proportioned embryos, each half the normal size, were generated within the space of a single embryo, is a perfect example of scaling (fig. 22; see also fig. 8).

Figure 22. How to adjust pattern with size ▶
Within a given species, different individuals vary in size. The relative size and position of limbs and organs must be adjusted, or scaled, to the global size of the organism. In molecular terms, the morphogen gradient across the tissue is adjusted to the field size.

Top: Wings of two fruit flies from the same species that were grown under different nutritional conditions and thus have different sizes. Regardless, the pattern of veins that results from signaling by a morphogen is perfectly scaled (D. Ben-Zvi and B. Shilo, Weizmann Institute); *bottom:* An example of scaling—the facial features of the large head sculpture and the girl display similar proportions (B. Shilo, Museum of Fine Arts, Boston)

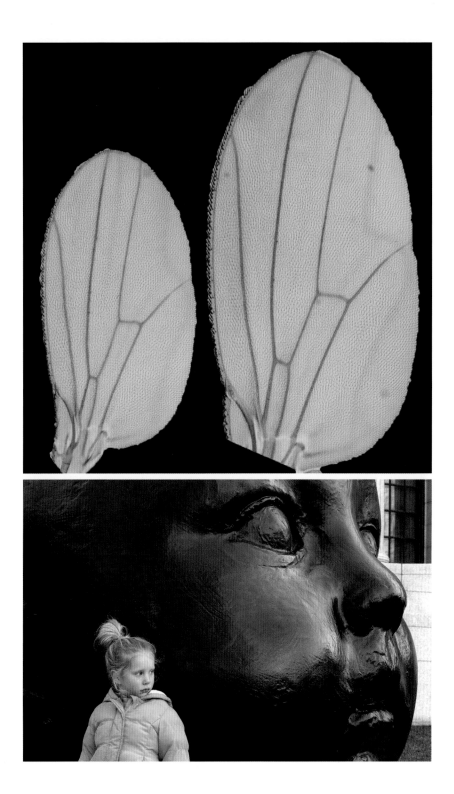

The poem "The Little Elf," by John Kendrick Bangs (1862–1922), conveys the relative aspect of scaling in a humorous way:

I met a little Elf-man, once,
Down where the lilies blow.
I asked him why he was so small
And why he didn't grow.

He slightly frowned, and with his eye
He looked me through and through.
"I'm quite as big for me," said he,
"As you are big for you."

Achieving this kind of proportional patterning is not trivial when we consider the morphogen model. A restricted group of cells produces the morphogen. It is then distributed in space by biological and physical rules that involve its diffusion rate, degradation, and uptake by cells it encounters along the way. But these rules are "local" in the sense that they don't relate to the global size of the embryo or organ; rather, they refer just to the immediate environment in which the morphogen is distributed. These local rules would be the same in large and small embryos. To achieve proportional patterning, an additional mechanism needs to be invoked.

Any mechanism that could account for scaling should involve two essential elements. First, the global size of the field should be measured. Second, the distribution properties of the morphogen should be adjusted, or scaled, to the magnitude of the field. In other words, some signal from the furthest edge of the organ or embryo should be able to "tell" the entire field that this is where the edge is located, and the shape of the morphogen distribution in the entire field should adjust accordingly. This tuning process is in essence what scaling means: fitting the shape of the entire structure to its relative size. How such a mechanism operates in practice is only beginning to become clear.

In the few cases where the molecular mechanism for scaling has been uncovered, it indeed involves a communication between the edge of the field and the machinery that controls the distribution of the morphogen. Danny Ben-Zvi and Naama Barkai have proposed a mechanism called expansion-repression. It involves two secreted molecules. The first component is the expander, which facilitates the spread of the morphogen. The second is the repressor, which is the morphogen itself. Imagine the generation of the morphogen gradient as a dynamic process in which the morphogen spreads from its site of production toward the edge of the field. Before it reaches the edge, the expander will continue to be expressed at the edge and will facilitate the morphogen's spread. Once sufficient levels of the morphogen reach the edge, they will terminate expression of the expander, and the morphogen pattern will stabilize. If the field is larger, it will take longer for the morphogen to reach the edge, and hence the expander will be expressed for a longer time, to facilitate the spreading of the morphogen profile. Exactly the opposite will happen in smaller fields.

In all the research I have discussed so far, biological experiments (in genetics, biochemistry, cell biology, and molecular biology) proved to be the basis for identifying the key elements guiding the processes of development, their molecular nature, and the circuitry they form. Analysis of problems such as scaling highlights a new trend in the study of developmental biology. The cells are actually sensing *the level* of the morphogen they encounter, and these levels have to be adjusted and calibrated to give rise to the correct patterning. We therefore need a quantitative understanding of the mechanisms guiding development.

Quantitative analysis poses a problem for biologists for two reasons. First, it becomes difficult, if not impossible, to measure such parameters within the intact animal. We would like to know the level of active morphogen at each position along the gradient. Often, however, the levels are both too low to be detected by available techniques and yet sufficiently high to elicit a biological response. Even if we can detect the morphogen, some of it may be in the active form, while the rest may be in an inactive form or in a cellular compartment where it is not functional. It is not obvious how to detect only the active morphogen form.

The second reason is that developmental biologists are trained to look at complete effects. These outcomes can be massive, such as loss of all head structures, or subtle, such as loss of a few hairs in the outer surface of the embryo. But biologists are less used to looking at quantitative effects within the whole organism. This is where close collaboration with scientists who are trained in quantitative analysis of data and theoretical modeling becomes extremely useful and synergistic. A viewpoint that is focused not on the molecular components of the system, but rather on mechanisms that could account for the desired or observed quantitative behavior of the biological circuitry being studied becomes essential.

When experimentalists and computational scientists complement one another, the outcome can be a deeper quantitative understanding of the behavior of the system. Such a collaboration also leads to finer insights into the molecular circuitry involved and the parameters that are critical for its quantitative output. "Hand waving" arguments of biologists are converted into equations that are being scrutinized. It is important to note, however, that computational approaches do not provide a magic wand that readily solves any biological problem. The task is to provide a new understanding, rather than a mere description of the biological system by a set of equations. Therefore, identifying the insightful computational solution that provides distinct and testable predictions of the system becomes as challenging as the biological experiments.

The need for quantitative understanding relates to the broader issue of robustness. On the one hand, we know that the structure of the embryo is formed in a reproducible fashion from one embryo to the next. On the other hand, the determination of its structure is based on communication among cells. And, as we saw in the case of the morphogen, this communication is exquisitely sensitive to the levels of the signal at any given point. This issue poses a problem in a biological system, because cells are subject to variability, which compromises their ability to provide a reproducible level of signal from one cell to the next and from one embryo to the next.

What could be the sources of such variability? Although all cells contain the same genetic material, they are not identical. Even two sibling cells generated by division of the mother cell may differ from each other. Suppose there are proteins that are present at low levels in the mother cell—for example, as low as ten molecules per cell. When the mother cell divides, there are many instances in which one daughter cell will contain six and the other four molecules, and so on. Other sources of variability could be different external temperatures or nutritional conditions in which the embryo develops.

A final source of variability stems from the number of functional gene copies. Each gene is normally present in two copies, one originating from the father and the other from the mother. In most cases, elimination of one of the copies for a given gene does not give rise to any detectable abnormalities, because the expression of the second copy is sufficient to provide the necessary function. But in fact, it is not clear why this is the case. We can assume that cells that have only one functional copy of a gene express half the level of protein as normal cells. How do cells get by with half of the normal level, especially in the case of proteins whose levels are critical, such as the morphogens we described?

There must be special mechanisms that are operating to counteract the quantitative fluctuations and provide a reproducible outcome. This is the essence of robustness. It entails a new level of quantitative regulation, which we are only beginning to understand. Essentially, it tries to explore how biological systems that typically operate and communicate in a "noisy" and variable environment can minimize the consequences of these fluctuations and provide a reproducible output.

Can Gradients Count Time?

In the classical morphogen gradient system, different cells in the field sense the morphogen at the same time. In reality, however, organs grow in size by cell division, in parallel to their patterning. If the source of the morphogen stays fixed over time, the implication is that the morphogen itself will display a different global pattern of spreading as the organ becomes larger. Earlier in this chapter I discussed scaling, which is the ability of a morphogen to adjust its distribution to the size of the organ at a discrete stage when the pattern is determined. Here I introduce another option, in which the successive growth of a tissue is also harnessed to determine the final pattern.

The five digits in our hands and feet are clearly different and distinct. The fate of each digit is determined at a time when the future limb (the limb bud) is small. A protein called Sonic Hedgehog, or Sonic for short, functions as the morphogen in this case. It is expressed in a highly restricted region, at the position of the future "pinkie." Because the limb bud is small at that stage, Sonic is secreted from the cells where it is produced, spreads to the extracellular space, and reaches all cells in the limb bud. At first, the different cells (including cells that do not produce it) are exposed to roughly similar levels of Sonic. As the limb bud grows by cell division, the cells that are farthest from the source of Sonic (representing the future thumb) become positioned outside its sphere of influence. Later on, the cells that will form the second digit (pointer) stop sensing Sonic, and so on. Once the growth of the limb bud is completed, different regions have experienced the influence of Sonic for different lengths of time, with the future pinkie being exposed continuously to the morphogen. By combining experimental and computational approaches, the laboratory of James Briscoe has uncovered the molecular mechanism by which the cells convert their history (that is, the length of time during which they are exposed to Sonic) to distinct cell fates. In other words, the morphogen exerts its effect as a summation of time rather than as a difference in concentrations that the cells experience at a given point in time (fig. 23).

This process is analogous to how long people are exposed to education. At first, everyone is exposed to the same compulsory system. But after compulsory education ends, the length of time someone remains within the educational framework will affect future career opportunities. We would expect people who have completed only elementary school to adopt a different set of professions than those who have graduated from college.

Figure 23. Using time to measure gradients ▶

Time can also be used to generate a graded signaling mechanism among cells. The fates of the digits in our arms and legs are shaped by the gradient of a protein termed Sonic hedgehog that emanates from the future position of the pinkie. Initially, when the limb bud is small, all cells sense Sonic. As growth ensues, cells gradually grow out of the Sonic domain. How long cells were exposed to this molecule will determine which digit they form. The cells exposed to the protein for the longest time will produce the pinkie.

Top: Ten-day-old mouse embryo, with the restricted source of Sonic RNA expression (purple) in the posterior part of each limb bud (J. Hu and C. Tabin, Harvard University Medical School); *bottom:* A flashing neon "Pizza" sign on Massachusetts Avenue in Cambridge. The letters appear successively, and thus, on a long exposure, the letter P appears brightest (P for pinkie . . .) (B. Shilo)

CHAPTER 7

How Do Patterns Evolve Rapidly?

The diversity of patterns in the organisms that surround us is astounding. Just contemplating the numerous species of birds or insects, each with its unique coloration and structure, is mind-boggling. In daily life, we regard this diversity as a static reflection of a variety that has always been out there. This assumption is, of course, a misconception. As brilliantly realized by Charles Darwin more than a hundred and fifty years ago, this diversity arose through evolution from a smaller number of species that diverged according to the rules of natural selection. Alterations in the genome are constantly taking place in a random, nondirected fashion. Most are silent, meaning that the resulting alteration in nucleotide sequence they generate has no effect on the expression pattern or structure of the proteins produced by the cells. Alterations that are not silent are usually harmful, and the organisms that carry them are wiped out. But in rare cases an alteration may give rise to a change that is advantageous to the organism in a particular setting and environment. This advantage may be translated to a higher survival rate of this organism and its descendants, such that eventually the change propagates and is stably carried by the entire population.

We might intuitively think that the evolution of new species and patterns is a lengthy process, but in fact dramatic changes can occur in just hundreds or thousands of years. The Galápagos Islands provide a living laboratory for evolution and were the first inspiration for Darwin's ideas. Far from mainland South America, these islands were formed by volcanic eruptions that took place 0.6 to 4 million years ago. From the perspective of the human life span, this may sound like a long time. When considered in the global time frame, beginning with the estimated appearance of the first cells on earth more than 3 billion years ago, however, this period is a tiny fraction, representing something like three seconds of an hour. The Galápagos Islands' many endemic species are a product of the evolution that occurred within this relatively short time span, including the flightless cormorant, the marine iguana, and giant tortoise species that are indigenous to one or another of the islands. Perhaps most famous are fifteen species of finches, each with a distinct beak form that maximizes their ability to feed efficiently in different habitats.

The evolution of patterns represents a strong and dynamic driving force. How does it take place in molecular and mechanistic terms? The answer had to wait until an understanding of the molecular mechanisms of development had emerged. Once these mechanisms became known, scientists could consider how cell communication pathways are used and modified to create new patterns.

If the genes that control patterns are conserved across evolution, how have they been modified to create the vast diversity of patterns in nature? The search for this answer led to the emergence of the field known as evolutionary-developmental biology, or evo-devo for short, for the study of how developmental patterns evolve. When we examine the development of structures in organisms that are not typically studied as model systems, such as butterflies, sea squirts, and hydra, we lack the research tools that are so powerful in studying fruit flies and nematode worms. These tools include the availability of mutations, the ability to manipulate the genome, and a detailed knowledge of the genes of these organisms.

This is where the enormous conservation of genes that guide development comes to the rescue. Even without being able to manipulate key genes in organisms as diverse as butterflies and sea squirts, we know where to start our molecular exploration. By examining genes that are shown to play key developmental roles in the organisms we have genetically scrutinized, we can try to determine where these genes are expressed in the new organisms under study. In many cases, they are expressed in similar places and are thus likely to fulfill analogous roles. In other cases, a distinct expression pattern of a gene provides an indication for its causal role in positioning this structure. For example, finding that a particular gene is expressed in a pattern that coincides with the appearance of pigmented spots on a butterfly wing suggests that it underlies their placement.

Beyond identifying the critical players in the development of patterns, how does the detailed study of development in an organism like the fruit fly help us understand the evolution of a pattern that generates new species? As we have seen, the number of pathways that guide the determination of patterns is limited; only a few are used during development. It follows that the same pathways are used reiteratively, tens and even hundreds of times, to comply with the large number of decisions needed to generate the enormous diversity of cell types that constitute the embryo and mature organism. Another feature of these pathways is their striking evolutionary conservation. We know that they have not changed in the past eight hundred million years of evolution because the components constituting these pathways are found in a broad range of multicellular organisms that have evolved over this time. This consistency indicates that any change in the structure of a protein within these pathways is lethal to the organism and will be excluded during selection in the course of evolution.

It is easy to conceptualize why the components of the cardinal signaling pathways preserve their structure throughout this wide evolutionary time scale. The pathways function as a "relay race," where one protein associates with and activates the next one. This association is based on structural complementarity, very much like a key that fits a lock. Without a matching change in the associated proteins, a single change in one component will block the pathway. Because the mutations that drive evolution are random, the chances of obtaining two such balancing mutations at the same time are extremely low. In addition, since the same pathway operates in numerous developmental junctions, the same key opens multiple doors. Thus, any alteration in the activity of a signaling protein will impinge on all the numerous junctions in which it works.

The preservation of the developmental pathways across the entire spectrum of multicellular organisms indicates that they were present in the primordial multicellular organism and were conserved as such to the present day in all organisms that descended from it. Although this ancestral organism does not exist today, we can imagine which patterning mechanisms it possessed based on the genes and pathways that have been conserved in all present multicellular organisms. It was quite an elaborate creature, displaying a segmented body pattern and with possible differences between segmented regions. This organism had appendages from either all the segments or some of them. It had primitive eyes that could sense light, and it probably had an immune system that provided protection from common infections by bacteria or fungi.

It is striking to note that in some cases not only the constituents of a signaling pathway but also the context in which it is used are completely preserved across wide evolutionary distances. One of the most compelling examples for functional conservation is the gene *eyeless/Pax6,* which dictates the position of the eye in all species. The restricted domain where this gene is expressed will define the position of the future eye. The protein is present in the cell nucleus and controls the expression of target genes. Its presence within the nuclei of the future eye cells provides the "eye context" to these cells, such that any signaling event that they will experience is interpreted within this context. Eventually, this protein leads to the expression of proteins that make the light-sensing machinery of the eye. The structure and organization of the eye in vertebrate and invertebrate organisms are very different because the proteins that constitute the eye differ. Yet the same protein controlling the expression of "eye-specific" genes in all cases implies that the circuitry has been conserved, even if the final players are distinct.

How can changes in the genome lead to alterations in pattern? One hypothesis that long prevailed was that genes duplicate in the course of evolution. Once two copies of the same gene are created, genetic variations will generate a distinct pattern for the expression of each gene and may also create specific changes in the protein structure encoded by each form. The broader view that we have today, stemming from the complete sequencing of genomes of many organisms across the evolutionary scale, tells us that this mechanism is the exception rather than the rule. Not only are the signaling pathways preserved, but the number of copies of each member is usually not altered. So where do the genetic changes take place to drive the evolution of pattern?

The basic structure of a gene can be divided into two distinct domains.

One part contains the information that will be transmitted through the production of RNA to the sequence of the protein formed according to the gene. Additional sequences on the DNA that surround the gene (described in Chapter 5 as landing pads) and can be positioned thousands of bases away from the coding information define the gene's expression program. In other words, they dictate in which cells and at what time the gene will produce RNA that will be translated into protein. These sequences are analogous to an oven timer that determines when and for how long the oven will be turned on.

The regulatory sequences of a gene constitute the landing pads and recruit the cellular machinery that will produce RNA from the template of the gene. These landing pads on the DNA can be positioned at a distance from the DNA sequence that is copied into RNA, since the DNA structure is flexible and capable of folding. Moreover, each gene can have more than one landing pad. In other words, a gene can be regulated by one set of proteins that occupies a particular landing pad in one setting and by a distinct set of proteins and landing pads in another. Although the coding sequence is unique, the gene can actually be viewed as a set of modules that can be independently regulated multiple times, by regulatory domains that are distinct from one another. Returning to the analogy of the oven timer, the same oven can carry out different cooking programs. Thus, without duplicating a gene, multiple patterns of expression of the same gene can be created simply by modulating the sequence of the landing pads.

What kind of changes in the sequence of the landing pads can lead to deviations in the expression pattern of a gene? Envision a gene that is a target of the morphogen-signaling pathway and is expressed in a region exposed to a particular concentration of the morphogen. Even a small change in the sequence of the landing pad of this gene that would facilitate tighter binding of the signaling protein to the DNA will allow the gene to sense effectively lower concentrations of the morphogen. Consequently, the gene will now be expressed at a broader region of the field.

The DNA sequence of the landing pads is particularly suitable for generating diversity. First, because the landing pads are insulated from one another, new patterns can be added or deleted in one without affecting the others. Second, changes in the expression pattern will not affect the sequence of the gene encoding the actual protein that will be made. Third, it is fairly easy to generate new binding sites by mutations in the DNA. Although the landing pads may direct the binding of several proteins to the DNA, each protein is attracted to a short sequence on the DNA, which is typically only six to ten nucleotides long. Furthermore, the binding site is not like a rigid lock that will accept only one key. Several variations in the DNA sequence can be recognized by the protein, and all facilitate binding, albeit some more effectively than others. We thus consider mutations in the regulatory sequences that define the landing pads as the prevalent mechanism to implement the evolution of pattern.

Changes in the timing or position in which cardinal developmental genes are expressed will lead to dramatic changes in pattern. Any alteration in the expression domain of the *eyeless/Pax6* gene will immediately result in a change in the global structure of the eye that will be formed. In a laboratory experiment, when this gene was expressed in the future wings and legs of the developing fruit fly, eyes were generated at these positions. The evolutionary changes we can envisage are less drastic. Instead, we can imagine changes in the regulatory region that would allow tighter binding of signaling proteins. This should lead to an expansion in the boundaries of the normal domain of *eyeless* expression and will immediately give rise to eyes that are larger in size.

Earlier, I mentioned the concept of hierarchical patterning (Chapter 5). Because development is a succession of simple decisions that build complexity, the earliest decisions have the most dramatic effect and impinge on the broadest regions. Changes in the expression of genes that are positioned at the top of the hierarchy will thus have the most pronounced effect on the final pattern. It is also important to note that a single change in the expression of such a gene is sufficient for generating a dramatic ultimate effect, because all subsequent responses follow the normal developmental paradigms. In other words, creating alterations in the pattern does not require simultaneous and complementary changes in several components. To conclude, there is a significant probability of implementing major changes through random mutations. These mutations affect when and where genes are expressed. If we consider our oven timer once again, even small alterations in the length or temperature of cooking a particular dish may significantly affect the outcome of the final product.

I will provide two more examples of dramatic alterations in pattern that

result from simple changes in the expression of key genes. The first concerns pigmentation patterns in *Drosophila*. An examination of the wings of two dozen *Drosophila* fly species that have diverged over the past sixty million years (out of more than a thousand described *Drosophila* species) reveals an amazing diversity of pigmentation patterns that serve to distinguish the species and facilitate courtship behavior within each species. The pigmentation is dictated by a defined expression pattern of genes that encode enzymes producing the pigment. The biochemical components of the pathway and the enzymes producing the pigment have not changed. The expression pattern of key and limiting enzymes, however, has been altered because of variations in their DNA landing pads, which has led to the corresponding changes in pigmentation. As I mentioned above, changes in tightness by which the signaling protein binds the landing pad can alter the size of the wing domain in which the gene is expressed. Alternatively, it is possible that regulatory sequences comprising different landing pads have been added or removed, such that the expression of the enzymes producing the pigment now takes place at totally different sites on the wing (fig. 24).

The second example concerns Darwin's finches. The initial notion of evolution was triggered in Darwin's mind when, as naturalist aboard the scientific expedition vessel HMS *Beagle*, he identified and described fifteen species of finches in the Galápagos Islands. These island finches differ primarily in

◀ *Figure 24. The evolution of pattern — changing the "notes"*

Patterns of organisms can be modified rapidly in evolution on exposure to environmental selection forces. But if the canonical signaling pathways are conserved and unaltered, how are patterns sculpted by evolution? Evolutionary selection does not work on the genetic information coding for the signaling proteins. Rather, it operates on parts of the genome that dictate "when and where" these genes will be expressed. Alteration in pattern follows, using the fixed rules of embryonic development.

Top: Wings of different *Drosophila* fly species that diverged over the past sixty million years. Although the pigmentation pattern is varied, it is achieved by alterations in the regulation of expression of the same proteins that synthesize the pigment (B. Prud'homme, N. Gompel, and S. Carroll, Institute of Biology and Development, Marseille, and University of Wisconsin and Howard Hughes Medical Institute); *bottom*: The same musician and instrument can produce a dazzling array of tunes, depending upon the notes chosen (B. Shilo, New England Conservatory, Boston)

the shape of their beaks, allowing them to access different niches in their environment for feeding. Some species have long beaks that permit them to extract worms out of cacti, whereas others have short, wide beaks that help them crush the hard seeds they pick up from the ground. Because speciation in the Galápagos Islands is relatively recent, how can we account for this dynamic change in beak shapes?

A group led by Clifford Tabin, a developmental biologist at Harvard Medical School, argued that it may be possible to find indicative changes in the expression pattern of key regulatory molecule(s) within the handful of developmental signaling pathways that mediate cell communication. A correlation between the magnitude of expression of a putative signaling molecule and the size of the beak would point to a causal relation. The group collected finch eggs and looked in the developing beak of the embryo at the expression of different secreted signaling molecules. Indeed, they found that the expression of one particular molecule, termed BMP4, correlated with beak size. Species with wide beaks had a broader pattern of its expression than species with long, narrow beaks. This was a compelling correlation, but could they be sure this expression is the actual cause of beak size? By implanting a bead that was soaked in this protein in the beak of a developing chick embryo, they observed that the size of the beak was indeed dramatically enlarged, thus implicating this molecule as the causal agent.

Evolutionary forces operate not only on the pathways that determine cell fate but also on the profile of genes that are expressed to implement these fates. Once the final cell fate is determined, each cell shifts to a second phase of patterning, involving the execution of its designated fate. This may involve the induction of tens and even hundreds of specific target genes, which manifest the actual cell fate and allow the tissue to execute its function. From a signaling pathway that acts in a sequential manner, passing information from one protein to another, the cell converts to a phase in which key tissue-specific proteins that are expressed turn on, in parallel, the expression of a multitude of target genes.

Why is this multitude of genes expressed in a coordinated manner in the same tissue and at the same time? These genes all have similar regulatory sequences, which can recruit the machinery that will produce RNA. In other words, the inclusion of a gene in the repertoire of genes expressed in a given tissue depends on the presence nearby of such regulatory sequences. In the course of evolution, by adding or deleting the regulatory sequences that generate landing pads and dictate gene expression in a given tissue, genes are added or deleted from the repertoire that is expressed by that cell type. These changes do not affect the global position or size of the tissue but rather its overall composition.

CHAPTER 8

How Are Cells Programmed to Follow Predictable Routes to Specification?

To this point I have stressed the concept that cells do not have an intrinsic program of what they will become. Instead, each cell receives instructions from its neighbors about which path to follow in developing into a given type of tissue or organ. This strategy operates in most multicellular organisms and is employed under conditions in which the embryo has an excess number of cells, so that cells that failed to receive instructions can be eliminated. This strategy also provides a safeguard against faults in the process, because if one cell fails to assume a particular fate, a neighboring cell can always replace it.

With these concepts in mind, I would now like to argue there are organisms that use a different strategy. As a general rule, it is important for the scientist not to be dogmatic about a paradigm that was discovered, even if it is a universal one, and be open to other biological solutions that take place. In the strategy for patterning that I present in this chapter, many of the instructions come from within the cell rather than the outside. As a consequence, the cells are preprogrammed to follow a distinct and reproducible route.

The organism in which this strategy has been explored in most detail is the nematode or roundworm, *Caenorhabditis elegans*. This organism serves today as a powerful model system in many laboratories, following the pioneering work of Sydney Brenner, who first realized the unique properties of this worm, and the wealth of novel biological concepts it can provide (he was awarded the Nobel Prize in 2002, together with John Sulston and Robert Horvitz). The adult roundworms live for several weeks and feed on bacteria in the soil or, in the lab, on bacteria plated on a culture dish. These tiny (about 1 mm long) organisms undergo a rapid life cycle: a fertilized egg grows into a fertile adult producing offspring in three to four days. The embryos develop in only fourteen hours, and by the end of embryogenesis a larva that is 0.25 mm long emerges. The remarkable aspect of this worm is that the larva always has 558 cells. It will then undergo a series of molts before forming the adult, which always has 959 cells. The small and fixed total number of cells already hints at a rigid developmental program. Indeed, when we look at a particular roundworm tissue, such as neurons, the number of cells is set at 302. Having a small number of cells does not mean that the developmental program is simple or that the genetic complexity is low. In fact, there are more genes in the genome of this roundworm (about twenty thousand) than in the genome of the fruit fly, even though the number of cells in the fly embryo is a hundred times greater than in the worm embryo.

How was the reproducible developmental program of the worm unraveled experimentally? The most important feature in these studies was the transparency of the worm at all stages. Thus, by focusing the microscope up and down, a researcher can observe all cells of the live worm. Because the cells have distinct shapes and positions, it is actually possible to follow each cell, even without employing methods to mark distinct cells. For several years and with immense patience, John Sulston followed the development of C. *elegans* embryos under the microscope. What was striking to him was the reproducible manner in which cells developed. In other words, after familiarizing himself with the system, he could forecast the fate of every cell in the embryo. This ability to predict cell fate indicated that there is a rigid underlying program.

An immediate outcome of this inflexible course is that cells of one type cannot replace a cell that was lost. It is possible to focus a laser beam on distinct individual cells within the developing embryo and kill them without affecting the neighboring cells. The first prediction, which was indeed borne out, is that the loss of distinct cells cannot be compensated. Second, it is possible to construct the entire lineage of cells by removing a particular cell and later asking which tissue is missing in the mature embryo.

This analysis resulted in a lineage map, tracing the descendants and ancestors of each cell in the roundworm with absolute certainty. It is similar to a human family tree in the sense that all members can be followed. It is, however, very different from a human family tree in that the lineage is completely reproducible from one family (worm in this case) to the next. As Tolstoy wrote in the opening sentence of *Anna Karenina*: "Happy families are all alike."

But even when the rigid developmental program is implemented in the roundworm, communication among cells still plays a significant role, and the canonical signaling pathways are operating here, too. They give the cells information about their general coordinates and neighbors and thus help to adjust and fine-tune the stricter developmental program that the cells are executing.

Having marveled at the ability to carry out such a predictable biological program, I would now like to explore the underlying molecular mechanisms that execute it with such precision. The major difference between the implementation of this program and the signaling programs we discussed earlier is that in the lineage strategy of patterning, the cues come from "within" the cells.

Let us consider a simple scenario: a cell is positioned within the embryo, so that it "knows" where the head and the tail of the embryo are, following signals it receives from neighboring cells. This cell is also dividing with a fixed orientation relative to this global axis. If the "instruction" is to position a critical substance only in the cell closer to the tail, one daughter cell will have the substance and the other will not. This difference in the composition of the two daughter cells can now drive them toward different cell fates (fig. 25). Although the asymmetry that is generated is internal to the cell, it relates to the external environment. Both the orientation of cell division and the axis of asymmetric segregation of components are aligned with the global external environment. Thus, while different cells independently generate distinct progeny, the overall structure of the embryo will be coordinated.

With this general principle in mind, we can examine its implementation along the different stages of development. First, how does the early embryo know the position of the future head and tail? In the worm, the site of sperm entry dictates the initial symmetry-breaking event, which is reminiscent of the mechanism identified in frogs. Whereas the distribution of components

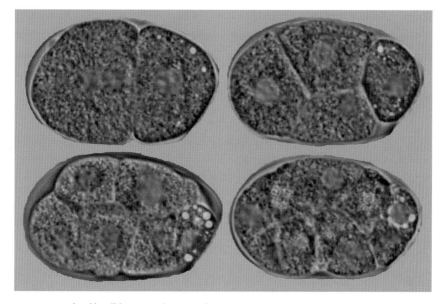

Figure 25. Predictable cell lineage in the nematode
In the embryo of the nematode, or roundworm, the shape and fate each cell will adopt
are completely predictable and generate a fixed family tree of cells, or lineage. After
sensing the general external coordinates of the embryo, intracellular components segregate asymmetrically between the two daughter cells at every round of cell division. The
distinct composition of each daughter cell will dictate its future cell fate. In this case,
globules termed P granules (green), which contain multiple RNAs and proteins that are
required to specify germ cells, are first segregated at the two-cell stage to the cell on the
tall side (right). At every subsequent division the P granules can be detected in only one
daughter cell, eventually marking the cell that will give rise to the germ line (D. Updike
and S. Strome, MD Anderson Cancer Center)

in the egg before fertilization is symmetrical, the random entry point of the
sperm leads to the segregation of components within the cell. This segregation results in the definition of a distinct axis, with the site of entry becoming
the future tail of the worm.

This asymmetry within the cell will determine the orientation of its first
division. Before every division, the cell contains a duplicated set of chromosomes that should be equally segregated between the two daughter cells. To
carry out the process faithfully, the cell generates molecular cables that are
built from a repeated unit. These cables are organized in parallel arrays and

are connected to opposite poles within the cell. Each chromosome is attached to these cables, which pull the chromosomes toward one of the poles of the cell. Each one of these poles will now give rise to a daughter cell. Thus, by orienting the cables, the position of the two daughter cells is determined. But why would the daughter cells be different from each other, and how can this difference be maintained and propagated?

To generate a stable difference between the two daughter cells, it is not enough to segregate a given component to only one cell. Eventually, this difference should be manifested in the expression of a different set of genes by each of the two cells. We learned earlier about landing pads on the DNA, which allow proteins to bind and promote the expression of target genes. Sometimes, these landing pads also promote the binding of proteins that inhibit gene expression, so that only when they are removed can gene expression proceed.

The asymmetric segregation of components between daughter cells in the worm embryo simultaneously affects the distribution of two elements. In the cell that will be activated, a protein that will drive gene expression accumulates. Concomitantly, another protein that inhibits expression is removed. In other words, in that cell a simultaneous push on the gas and release of the brake takes place. What is the advantage of this dual system? It is believed to be more reliable and rapid, because the final outcome depends not only on the absolute level of the activator or the absence of the inhibitor but also on the ratio between them. Thus, even when the inhibitor is not completely removed, there will be a sufficient excess of the activator to drive the process forward. In the analogy used previously, the car can go forward when you push hard on the gas, even if you still press very lightly on the brake.

The other cell, which is not activated, will maintain the inhibitory brake and follow an alternative program by default. In other words, there is a dictated course of events that takes place when no activation signal is provided. Thus, the asymmetric segregation of components has to push one of the cells in the desired direction, whereas the progression of the cell that did not receive the signal is set on "autopilot." With the availability of the default course, essentially only one active decision is made, yet the outcome displays two distinct daughter cells. It is important to emphasize that both courses represent dynamic programs of gene expression that progress in parallel, even though only one requires an active trigger.

The same biasing mechanism is used repeatedly for most of the cell divisions in C. *elegans*. How does the same biasing mechanism generate the diversity of cell fates? Here we encounter a concept similar to the one described in Chapter 5 for the reiterative use of the signaling pathways among cells. As a cell assumes a particular fate following the first division, it changes its internal composition of proteins. When its daughter cell encounters the same biasing mechanism in the next round of division, that cell will express a different set of target genes because of its altered internal composition. Thus, successive deployment of the same biasing system, coupled with a consecutive change in cell composition, can drive cells toward increasingly specific fates.

Although the nematode uses predominantly internal cellular mechanisms that bias the distribution of critical components within the cell, communication among cells through the conserved signaling pathways is also employed on numerous occasions. Localized activity of signaling pathways among cells sharpens the asymmetric segregation of critical intracellular components that eventually dictate distinct fates to the two daughter cells. In view of the smaller size of the worm embryo and the more spatially restricted cell fate choices that take place, the communication among cells in the worm takes place over shorter distances and affects only a limited number of cells at each time, in contrast to the large domains that are patterned by cell communication in other organisms.

CHAPTER 9

Ensuring That Embryo Development Is on the Right Track

Embryogenesis is completely different from an industrial production line on which preexisting parts are assembled according to an instruction manual. It also bears no resemblance to the casting of identical sculptures using a fixed mold. We should regard each embryo as a novel creation, guided by numerous and successive events of communication among cells. Perhaps a good analogy would be the production of multiple wheel-thrown ceramic bowls. Each bowl is made anew, but if the ceramics artist is skilled, the same motions can be accurately repeated to generate similar, although not necessarily identical, bowls (fig. 26). With this view in mind, it becomes even more compelling to explore how embryos of each species are made in such a reproducible manner.

Checks and Balances

Embryogenesis progresses from one step to the next, with no external inspector to ensure that processes are on the right track and to adjust them when coordination is lost. Instead, all the controls are built into the system and function from within. This feature requires that every step will be carried out with the utmost accuracy to prevent mistakes and imperfections. The need for tight control is even more evident in situations of morphogen signaling, in which the signals are not just on-off commands but also give rise to different outcomes depending on the signal level. How are checks and balances built into processes that depend on biological constituents and, as such, are subject to fluctuations and inaccuracies?

The main idea is that each process should control its own precision. The key principle is feedback, in which the product of a biological process adjusts its own production. As an example, consider the thermostat used for heating a house. When the temperature rises above a designated level, the thermostat shuts off the heat. Later, after the house cools and the temperature drops below the same value, heating will resume. So the temperature at any given moment (the final product) will influence the heating process by turning the heater on or off, depending on the thermostat setting.

The most common biological equivalent is a system known as a negative feedback loop. The activation of a given pathway leads to the accumulation of a substance, which attenuates the same pathway by which it was produced. This prevents the activity of the pathway from rising continuously and maintains it at the desired level. Attenuation of the pathway can be obtained by the final product of the pathway or by synthesis of an inhibitory molecule that is dedicated to the regulation of the pathway. The principle behind the inhibitory molecules is that they associate tightly with a component of the pathway and in so doing prevent this element from carrying out its function. In other words, the inhibitor forms a nonfunctional complex. For example, an inhibitory molecule that traps the protein that activates the receptor will reduce the level of signaling among cells.

Figure 26. Generating reproducible patterns   ▶

How are patterns generated reproducibly by cell signaling, despite variations and "noise"? This precision is achieved by intricate control mechanisms that modulate the signals transmitted among cells and "filter" out noise such that the final outcome is invariable.

Top: The hairs at the base of the fruit fly wing consist of several types and form an intricate array. When we examine similar organs from six different flies of the same species, it is striking how reproducible and stereotyped these structures are. The number and types of hairs and their position and spacing are virtually identical from one fly to the next (T. Kornberg, University of California, San Francisco); *bottom:* A skilled ceramics artist can generate reproducible shapes of wheel-thrown bowls. Like the process of embryonic development, there is no "mold" that will produce identical patterns. Every bowl is generated anew, but the fidelity of the procedure assures the accuracy and reproducibility of the end products (B. Shilo, Mudflat, Somerville, Massachusetts)

In many cases, the inhibitory molecules are ancient evolutionary relatives of functional molecules in the pathway. They have maintained their association with a given protein but lost the biological output that is triggered by this association. To illustrate the principle, think of the cord that connects the cell phone to the computer. If you cut this cord so that it has only the end that connects to the phone, it will function as an inhibitor by blocking use of the socket that is normally used for connecting to the computer.

Other types of control mechanisms have also been identified. An urban legend maintains that two independent keys are required to trigger an atomic bomb. Similarly, in many developmental junctions, the simultaneous activity of two inputs is required. This requirement prevents the false activation of a critical biological process and ensures that the pathway is activated only when two independent signals are present at the same time and place.

Additional regulatory strategies control the time element. How do you build a biological circuit in which a delay in the response to an initial signal is implemented? One strategy is to have the initial signal trigger a second response. Only when the first and second signals are simultaneously presented will the required process occur. The delay is generated by the time required for the first cue to create the second signal.

As these examples imply, different combinations and strategies can provide the desired outcomes in a reproducible and regulated fashion. Conversely, different regulatory wiring of the same elements can create distinct outcomes. Beyond identifying the components of a pathway, scientists are investing significant research efforts in elucidating the regulatory logic. Ultimately, elaborate and reproducible outcomes rely on the intricate regulation of a fixed number of operating components.

This discussion relates to a situation in which a given circuit is the sole regulator of a biological outcome and provides the necessary safeguards to protect the system from inaccuracies. Another strategy, redundancy, is often used to compensate, in a different way, for the variability and occasional unreliability of biological circuits. The simplest intuitive way to explain redundancy is that "two are better than one." When constructing a mechanical system that will carry out vital functions, engineers always design and install a parallel backup system if there is even a remote chance of system failure. This is what keeps most of us calm when boarding an airplane that will take us to

Figure 27. Backup mechanisms
If one of the trumpet players misses a few notes, the other players will back him up, and the overall tune the band is playing will not be compromised. Similarly, in the developing embryo there are many cases of genes that carry out overlapping functions, such that defects in one gene will be covered by the activity of the others (B. Shilo, Honk! Festival of Activist Street Bands, Somerville, Massachusetts)

an altitude of thirty thousand feet. Similar backup circuits are employed in many biological systems.

In simple terms, biological backup systems are employed when two genes carry out overlapping functions. Thus, when the activity of one gene is compromised, the backup gene will take over (fig. 27). The backup gene is often similar in structure to the original gene and arose in evolution by a process of gene duplication. This is distinct from the normal gene replication that occurs at every cell division, in which all the genetic information is copied accurately to generate a new set of chromosomes for the daughter cell.

Here, a single gene has been duplicated, and now the genome stably carries two copies of the gene. Each copy can evolve independently with respect to both its coding capacity and its regulation.

While such backup mechanisms provide biological robustness, they are a source of headache and complication for genetic research. The essence of genetic approaches is to study the function of the gene based on its absence. From processes that go awry we can infer how the missing gene normally functions. In the case of gene redundancy, in which a backup gene covers up for the other, eliminating just one of the two genes may produce no abnormalities or only very subtle ones. To identify the complete role of a particular circuitry, both genes need to be eliminated in parallel.

When Patterning Pathways Go Awry

In spite of all the elaborate regulatory systems just described, processes can and do go wrong, leading to defects and aberrations. When we study these defects in model organisms, we unravel the mysteries of the pathways and their functions. And because these same pathways play central roles in all vertebrate organisms, similar defects can also lead to human disease. In this respect, the identification of the developmental pathways in simpler model organisms can actually be an effective tool for identifying the genetic basis of human pathological abnormalities.

One class of pathological situations involves external environmental insults that disrupt the normal process of development. Cyclopia is a lethal malformation in which a single central eye is formed, brain hemispheres are incompletely developed, and most jaw and nasal structures are absent. This phenomenon is not just a term in Greek mythology. In a population of sheep in Utah, up to 6 percent of newborn lambs displayed these defects. On investigation, the culprit was identified as the California corn lily or California false hellebore (*Veratrum californicum*), on which pregnant ewes had grazed. When they ingested this plant on about the fourteenth day of gestation, they gave birth to lambs with birth defects such as cyclopia. The plant toxins were identified as the teratogenic chemicals jervine and cyclopamine.

Having identified the causative agent, the researchers next wanted to know which biological mechanism was being disturbed. The severe defects suggested that a pathway affecting the formation of brain and facial structures was compromised. One of the handful of pathways identified in the genetic screen as a major player in patterning the fruit fly embryo is the Hedgehog signaling pathway (see also Chapter 6 for a discussion about the fate of vertebrate digits). The *hedgehog* gene encodes a protein that triggers a signaling pathway. Like other developmental pathways, the Hedgehog protein associates with a receptor on the membrane leading to transmission of the signal to the nucleus. The name hedgehog comes from the appearance of fly embryos deficient in elements of this pathway. When Hedgehog is absent, the fruit fly embryo, instead of displaying hairs on the belly side in only one part of each segment, produces a lawn of thick hairs along the entire belly side, thus resembling a hedgehog.

The Hedgehog signaling pathway is highly conserved in evolution. Because the genome of vertebrates has undergone a series of gene duplications over time, today there are three genes codings for this protein instead of just one. When researchers isolated these genes in 1993, they wanted to name them by maintaining the relation to the original fly mutation name. There are several hedgehog species, so they named one of the genes Desert hedgehog and the second one Indian hedgehog. When it came to naming the third gene, however, "Common European hedgehog" just didn't roll off the tongue. Bob Riddle, a post-doc in Clifford Tabin's lab working on the project, came across a children's coloring book featuring the character Sonic the hedgehog (before the popular video game was marketed in the United States). Since Bob also played in a rock band, this name seemed to fit the bill, and the gene was named Sonic hedgehog.

This gene turned out to be the most versatile and central of the three. As the more than 3,600 papers in scientific journals all dealing with the gene attest, Sonic hedgehog plays a key role in numerous developmental processes of vertebrates. And this brings us back to the one-eyed sheep embryos. Sonic hedgehog is expressed in the developing "floor plate," which patterns the nerve cord and leads to its subdivision into the left and right brain hemispheres and eye field. Cyclopamine, one of the toxic plant substances eaten by the ewes, interferes with the transfer of signals through the Hedgehog pathway by blocking part of the receptor. In the developing sheep embryo, the attenuation of signaling through this pathway gives rise to cyclopia.

Another (in)famous example of an adverse chemical agent is thalidomide, which gives rise to the aberrant development of limbs. Originally developed in the late 1950s and early 1960s to treat nausea in pregnancy, this drug induced severe pattern abnormalities in fetuses. We now know more about the molecular pathways operating during limb morphogenesis and how thalidomide impinges on them. Fetal alcohol syndrome represents another example of toxic effects at the critical stages of human gestation and can result in the malformation of the brain and facial structures, known as holoprosencephaly.

A second class of patterning defects is caused by mutations that alter the structure and subsequent activity of critical components in developmental signaling cascades. A normal organism over its lifetime can acquire mutations within some of its cells. A mutation will appear in a single cell and will be transmitted to that cell's progeny. These are known as somatic mutations and are not transmitted to the organism's offspring. It is important to note that the developmental pathways not only operate during embryogenesis but also control cell growth in the mature organism. Because somatic mutations may occur late in the life cycle of the organism after normal embryogenesis has been completed, they will not affect patterning. But if a critical control in the proliferation of adult tissues goes awry, pathological situations involving cancer may arise.

The most common type of human skin cancer that occurs predominantly in the exposed head and neck is basal cell carcinoma. This condition is also related to the Hedgehog pathway. A mutation in the receptor for Sonic hedgehog causes too much activity, even in situations when Sonic hedgehog is not present. In other words, the pathway is short-circuited and is activated constantly, regardless of external inputs. This situation represents the opposite scenario to what we observed with cyclopamine, which inhibited the activity of the receptor.

The skin contains cells that proliferate and continuously give rise to differentiated skin cells. Normally, the process is highly regulated, so that the newly generated cells differentiate and replace the dying cells, thus maintaining a constant number of skin cells. The process relies on stem cells (see Chapter 11). Perpetual activation of the Hedgehog pathway in these cells can tip this fine balance and lead to excess cell proliferation without the accompanying process of differentiation. The result is a dramatic increase in the number of proliferating cells, giving rise to a tumor.

One treatment for basal cell carcinoma involves artificial attenuation of the Hedgehog pathway to alleviate the circuit that is continuously "on." Cyclopamine could provide this inhibitory activity. This is another example of the diverse approaches used in the study of a universal signaling pathways. With scenarios that are as disparate as grazing sheep and human cancer, seemingly unrelated stories suddenly come together in the most compelling way when a common pathway is implicated.

One of the most prevalent ways to regulate the proliferation of cells in the body involves a signaling pathway termed receptor tyrosine kinases. This pathway also employs activating proteins, membrane receptors, and a cascade of proteins that sends the information to the cell nucleus. Because this pathway has a central role in the regulation of cell growth, elements of it are prevalent targets for mutations that give rise to a variety of human cancers. This pathway can be short-circuited by mutations that produce a constantly active version either of the receptor itself or of elements that mediate the signal between the receptor and the nucleus. Another strategy that is often used by cancer cells is to increase the number of gene copies in the genome that encode the receptor. The resulting cells will produce higher levels of the receptor. With these amplified "antennas," the cancerous cells are much more sensitive to the normal levels of activating proteins they encounter and proliferate under conditions where normal cells would stop growing.

When cancer cells produce more receptor belonging to the receptor tyrosine kinase family, this increase serves as one of the most reliable prognostic predictors for the stage of a tumor, especially in breast cancer. In other words, high levels of the receptor indicate that the tumor is more aggressive. This indicator also presents a window of opportunity, however, because it provides a way to tackle the tumor cells specifically. By injecting antibodies that bind to the portion of the amplified receptor that is protruding outside the cell membrane, it is possible to specifically attack the tumor cells and, in combination with chemotherapy, provide an effective treatment.

In Chapter 1, I described my quest for genes in simpler organisms that would be similar to the human cancer-causing genes. One of the genes I used in this search is a central component in the transmission of information from receptor tyrosine kinases to the nucleus, and this gene is mutated in about 25 percent of all human cancers. Indeed, I identified a similar gene in the genome of the fruit fly. When we finally isolated the fruit fly gene and determined its sequence, it turned out to be 75 percent identical to the encoded sequence of amino acids in humans. The degree of similarity is so high that a gene from one species can be used to replace the activity of its counterpart in the other species, in spite of their evolutionary divergence of over six hundred million years! For example, we inserted the human cancerous mutation into the sequence of the fly gene, and this modified gene was incorporated into the genome of mouse cells. When we injected these cells into mice, aggressive tumors were formed that were indistinguishable from the ones caused by the homologous mouse or human cancer-causing gene.

Over the years, research work has identified a host of developmental junctions that are dependent on pathways triggered by receptor tyrosine kinases. In the roundworm, the determination of the organ that produces the eggs (the vulva) relies on a single cell that provides the activating protein, and the adjacent cells that respond to it produce the vulva. In the fruit fly, dozens of events rely on this pathway. They include patterning of the skin cells that form the belly, the nervous system, the eyes, wings, and legs. Although it may sound obscure and surreal, by studying the development of the compound eye of the fly or the vulva of the worm, we have made enormous progress toward understanding the basis of human cancer. The primary motivation of researchers studying these processes in the model organism was the desire to understand how organs are formed in the model systems they used. Yet their findings have immediate bearings on human health and disease.

A critical component in controlling the number of cells relies not only on when cells divide but also on the elimination of cells through apoptosis, or programmed cell death. Interestingly, the term was adapted from a Greek word that refers to leaves falling from trees or petals falling from flowers. This process is used in many adult tissues as well as in the developing organism. Excess cells are removed in an orderly manner during development to generate the final pattern. In fact, this killing is a central component in the overall strategy of development. Generally, the organism harbors an excess number of undifferentiated cells. Through the processes of cell interactions, cells assume the fates needed to build an eye or a nervous system. But in the end, so that a perfectly streamlined organ is constructed, the excess cells that did not contribute to the final structure are eliminated and removed. In other words, being on the "safe side" of having extra cells also requires a mechanism to remove such cells at the end of the process. The controlled killing of cells contributes to the normal balance of cell numbers. Abnormalities in this process can also lead to an excess number of cells and unregulated growth of a tissue, giving rise to tumors.

Because the regulated removal of cells is essential for the final structure of an organ, genetic approaches that monitor defects arising from the absence of such cell removal can be very powerful in dissecting the underlying mechanism of programmed cell death. Quests for mutations that impede the selective removal of very specific cells in the worm, pioneered by Robert Horvitz, have identified the cardinal elements in the killing pathway and its regulatory logic. The tightly regulated removal of cells—131 cells to be precise—is what contributes to the final magic number of exactly 959 cells in each worm. Indeed, many of the elements of this cell-killing pathway that we know today and are used in a wide range of organisms from worms and flies to humans were uncovered by such genetic searches.

How Worms and Flies Help Combat Human Disease

Information about the complete, or almost complete, sequence of the human genome became available about a decade ago. Continuous innovations in the technologies that are used for obtaining a complete sequence of a human genome and assembling it are giving rise to a process that can only be described as a technological revolution. In the foreseeable future, it should be possible for an individual to readily obtain a complete sequence of his or her genome for the price of a weekend at a luxury hotel. This capability comes with a baggage of moral and ethical issues that I will not touch on here.

For the purpose of our discussion, I note that this revolution allows rapid identification of genes that cause a variety of human diseases. We will shortly find ourselves in a situation where the nature and sequence of many genes that underlie a variety of human diseases will be known. This is more definitive when a single gene is the source of an abnormality, as in cystic fibrosis. The analysis becomes much more complicated for diseases that involve simultaneous changes in more than one gene, and where different combinations of gene abnormalities can cooperate to give rise to a similar syndrome, as is the case with autism or autoimmune diseases.

I have expounded on the ability to identify the central players that control cell growth and death in humans through genetic analysis of developmental processes in worms and flies. The role of these organisms in combating human disease does not end there, however. Genetic approaches in model organisms also provide an effective and versatile platform to develop new therapeutic approaches targeted for humans. Often, when the gene underlying a certain human disease is uncovered, it is not obvious what molecular process is driven by the protein that it encodes and how that relates to the eventual manifestation of the disease. Such an understanding is conceivable if additional proteins that participate in the same biological process can be identified to provide a clue to the general mechanism.

We have discovered that more than half of the genes implicated in human disease have counterparts in roundworm and fly genomes, which can be identified by virtue of their sequence similarity. It is therefore logical to assume that the proteins with which they interact will also resemble the assemblage used in humans. If we start from a gene with an unknown function, how can we identify its partners and uncover the biological process they regulate?

One approach is based on the biochemistry of proteins. In parallel to the revolution in DNA technology, there have been great strides in analyzing proteins and protein complexes. In many cases, proteins that function together are not only present at the same place and time but also stick to one another. This association can be transient, for example, in cases where they briefly "touch" to transmit a signal. Associations among proteins can also be more stable, such as when groups of different proteins together generate a firm complex.

If we treat one protein in a complex as "bait," then isolating that protein can help us identify the entire complex of proteins it associates with. The incredible technologies that are now available allow us to take this mixture of unknown proteins and fragment them into smaller pieces. We can then reveal the identity of the proteins by taking extremely precise measurements of the size of the fragments. This procedure is possible because the complete sequence of the genome allows us to predict the size of fragments that comprise every protein that can be produced. So in cases where the protein implicated in a disease associates with other proteins, we can identify these partners molecularly.

Another approach to identifying genes in the same pathway employs the workhorses of genetics: fruit flies and roundworms. I previously described the enormous utility of genetics when a gene is removed and the consequences are monitored. It is also possible to interrupt a signaling pathway more subtly, by reducing the activity of a gene or, conversely, by elevating it above the normal level. In this case, a much less dramatic effect can be observed. For example, I mentioned the genes that cause cancer and are involved in a variety of developmental switches in model organisms. When these genes are removed, embryos die at early stages of development. But when their activity is merely modulated, it is possible to observe, for example, the subtle effect of disarray in the compound eye of the adult fly. This result is achieved because all the processes this gene regulates are not completely blocked, and particular processes may become more sensitive to the change in activity.

After we have generated a tissue that is "sensitized" to the activity of a given gene, we can use genetic means to identify interacting genes. Recall that every gene is represented in the genome in two copies. When we remove one copy of a gene, we halve the level of the protein it will make, so that the gene product remains, but in a lower amount. For most processes this would indeed not matter, but when a pathway is sensitized, having half of the level may make a difference. Think about hungry diners waiting for their lunch orders at a bustling, busy restaurant. They must wait not only for the kitchen staff to cook and plate the food but also for their server to bring them their dishes.

After we have identified a disease-causing gene, along with other genes that operate in the same pathway, our next challenge is to modulate and correct the activity of this pathway. If the disease is caused by overactivation of the pathway, as in the case of cancer, then we search for and apply drugs aimed at attenuating the activity. When defects result from reduced activity, the goal is to boost the pathway's activity or to remove proteins that normally restrain it.

I mentioned previously that cancer-causing receptors can be inhibited at the external surface of the cell. But most of the drug targets are actually located within the cell. To be effective, a drug has to be able to cross the barrier of the cell membrane and be targeted to its site of action. The requirements for such drugs are stringent. They need to trigger the desired response in a specific manner, so that side effects caused by fortuitous interactions with other pathways, as well as interactions with the relevant pathway in normal cells, are minimal.

The preferred methodology for designing drugs is to create small synthetic chemical molecules that can readily cross the cell membrane. These molecules complement in structure the site of the target protein that should be modulated. The higher the complementarity is between the drug and the protein, the tighter the binding. A snug fit will naturally obstruct the site on the protein from performing its function. Although the assessment of fit between the two can be carried out in solution in a test tube, unforeseen factors within the cell and organism always impinge on the final outcome. Fruit flies and roundworms provide an efficient testing ground for such chemical compounds. If a specific biological assay is available—for example, alteration in eye pattern following expression of high levels of the tested protein—then the effect of a large number of small chemical molecules can be readily tested in order to identify the best drug or combination of drugs.

CHAPTER 10

Shaping the Tissues

In previous chapters I have described the decisions that lead cells to become one type or another. These decisions are mediated by a small number of communication pathways that are used repeatedly at each decision junction. Making a developmental decision is akin to buying a plane ticket, in the sense that the cell has chosen a future path but has not yet embarked on the journey and executed it. However, in contrast to the limited number of genes involved at each decision junction, the actual execution of the cell's fate may require the activation of hundreds of genes and proteins.

Cell fate is eventually executed at the level of individual cells, which assume their distinct structures and express the proteins that will enable them to carry out their function perfectly (fig. 28). Consider, for example, the components of a nerve cell. To adopt its typical organization and carry out its intricate roles, hundreds of proteins that are specific to the nerve cell must be expressed. These include proteins that maintain the elongated shape of the axon that extends from the cell body to distances of tens of centimeters, proteins that transport cellular components along these long extensions, and proteins termed channels, which conduct the flow of charged ions through the membrane.

Execution also takes place at the level of groups of cells, which together shape distinct tissues and organs. Cells continue to communicate and coordinate after the decision to become a particular cell type. The actual building and elaboration of tissues continue to require tight synchronization but also involve other molecular means of communication and coordination. Let us outline some of these mechanisms now.

Cell Migration

Embryogenesis is a highly dynamic process. Cells not only divide and acquire distinct fates throughout the process but also move continuously. Cell movement, like cell fate determination, is highly regulated and as such plays a central role in shaping the final structure of the embryo and its tissues.

Cells migrate for several reasons and purposes. The most obvious is that a particular cell type may be defined at one location of the embryo but is actually required for the formation or function of another region. Think of goods such as food or clothing that are produced at a central factory and then distributed to the diverse locations where they are needed and used.

How does a cell migrate? To move in a particular direction, the cell must carry out a chain of coordinated processes. It is important to note that cells are not freely hanging in space but are tightly associated with either the substratum or with neighboring cells. These contacts are necessary to hold cells in place when there is no movement and to provide traction forces when movement does take place. To initiate movement, cells that are crowded and tightly associated with other cells must first detach from their neighbors. Movement of individual cells begins with the stretching of the cell in a certain direction. After extensions have been generated in the direction of movement, then selective detachment at the opposite end of the cell takes place. Finally, the detached region follows the leading end, as the cell itself

Figure 28. Cell shape is adjusted to function ▶
Once the directions for the final fate of a cell are delivered, the cell must now execute these instructions. Each cell expresses numerous proteins that will manifest its differentiated character and shape it to fulfill its ultimate function.

Top: Vestibular cell of the mouse inner ear, which provides the sense of balance. Movement of hairlike extensions by the surrounding fluid conveys spatial orientation (A. A. Dror and K. B. Avraham, Tel Aviv University); *bottom:* Tools, such as the wood plane, are crafted to execute their specific function most efficiently (B. Shilo, Center for Adult Education, Cambridge, Massachusetts)

exerts a pulling force on its different parts to move the cell body and the internal organelles in the direction of migration, and the entire cell has thus moved from one point to another. Migration of a cell is analogous to a person confined within a sleeping bag trying to move with the bag in a particular direction. First an extension in one direction is generated, and subsequently the trailing end is lifted and advanced.

Both the initial extension and the pulling forces rely on an internal cytoskeleton of the cell. Critical components in the cytoskeleton are cables or microfilaments, which are formed like adjoining Lego blocks from the assembly of repeated units. These protein units are termed actin. Because these filaments are rigid, their elongation provides the forces that push the cell membrane forward. Think of a Lego tower that is covered by a sheet of cloth. Every time a new cube is added to the top of the tower, the sheet will be pushed forward.

The same filaments also guide the movement of subcellular structures in a directional manner. How is this directionality executed? Special motor proteins called myosins are attached to a cable of actin at one of their ends and are associated with the cargo that needs to be transported at the other. The cargo advances as the motors migrate along the actin cables, just like cars of a train that are linked to an engine that moves along the tracks. The motors use energy that is stored within the cell to fuel their movement.

When migration is coordinated among cells, even very short routes of individual cells can lead to dramatic effects on the overall pattern. For example, if each cell in an interdigitated, or interlocked, layer moves a distance of only one cell row in the same direction, a dramatic alteration of the entire layer takes place. Such a coordinated short-range movement is called convergence-extension, because the convergence of cell rows immediately leads to an extension in the global structure of the tissue. For example, during embryogenesis the round shape of the amphibian embryo becomes long and narrow. This elongation happens by coordinated movement of cells, in which cells from two adjacent rows slide to form a single row. Obviously, every row of cells becomes longer and narrower, and the combined effect of all the rows now leads to a similar effect on the embryo at large. Think of two rows of people standing in line in front of two ticket booths. If one of these booths closes, the two lines have to merge into a single line to purchase a ticket. Although the only movement necessary is the convergence of two rows, this merge can lead to a dramatic change in the pattern of the queue, which becomes narrow but twice as long.

Cells can migrate individually—white blood cells, for example, are attracted to sites of inflammation. The most prevalent mode of migration in embryogenesis, however, is collective cell migration, in which groups of cells migrate in a coordinated manner. This mode of migration allows cells to shape tissues and organs. Collective migration keeps the overall tissue structure intact during remodeling. In collective migration, mobile cells can transport immobile cells along with them. The integration of information transmitted among the migrating cells, combined with signals received from neighboring cells and tissues, provides an important layer of elaboration and refinement.

The information for the direction of collective migration is not stored in the migrating cells themselves, although their organization can facilitate directional migration. Consider a layer of cells that originates from a defined source and are all associated. Although the cells may all be identical, those at the edge farthest from the source differ from the others because they are not connected to neighboring cells at one end. If the cells are positioned on a molecular surface that is conducive for migration, the cells at the edge will provide the pulling forces that drag all the other cells toward the free edge and away from the cell source.

How to define molecular migration tracks becomes especially important when long-range migration takes place. In many cases, the external guiding cues provide much more than just a linear track. The external direction signals are usually secreted proteins that activate receptors on the membrane of the migrating cells. The interaction between the signal and the receptor within the migrating cells will trigger processes that define the position in which the actin microfilaments are formed, thereby causing the cell to extend in the desired direction.

Because the cells usually migrate on a landscape that is highly patterned, this background is capable of providing tightly coordinated and stereotyped signals for migration. Instead of picturing a continuous track, we can view such signals as dynamic signposts. These signals are similar to a car with flashing lights that guides a just-landed airplane along the maze of airport tracks to its final gate. The biological signals are dynamic, in that they appear first at one position and later at a more advanced position, with the migrating cells following suit.

The identity of the migrating cells and the interactions among them lead to the formation of elaborate yet reproducible organ structures. This result occurs in combination with the interactions with neighboring nonmigratory cells, which provide the refined cues for migration. A stereotyped collective migration that is guided along tracks will ultimately shape the structure of the organ. The final elaborate organ structure embodies the history of migration of the tissue (fig. 29).

Physical Forces

The rules of the chemical and physical world impose constraints on all cells. Chemical directions of interaction among molecules, based on either proximity or electrical charges, determine how proteins fold within the cell, impinging on their three-dimensional structure and subsequent function. Physical forces operate within a larger setting, at the level of compartments within the cell or between cells.

A crucial feature of cells that allows them to maintain viability and function is the ability to exert forces that can transport subcellular structures from one place to another within the cell and to move whole cells. I have noted the use of motors that move on actin cables and translocate their cargo within the cell. Their activity provides a molecular means to generate physi-

Figure 29. Cell migration ▶

Processes of regulated cell migration sculpt organs. Cells and tissues may end up in a body region that is distinct from where they originated; the migration path will determine the final shape of the organ.

Top: Cells that will form the tubular tracheal system of the fruit fly embryo (brown) are migrating in a stereotyped manner. Tracheal tubes formed at each segment are joined to form a continuous network. Migrating tracheal cells respond to dynamic spatial cues presented to them by neighboring nontracheal cells at similar positions for each segment. The similar migration program in each segment will give rise to the repeated structure of the trachea. Mature trachea will transfer air directly to all tissues (B. Shilo, Weizmann Institute); *bottom:* Sheep guided by a Bedouin shepherd. Like the migrating cells, the sheep follow a dynamic directional cue (B. Shilo, Rehovot outskirts, Israel)

cal forces. I also have mentioned that the addition of units to these cables generates a force that can push the membrane of the cell forward and initiate cell migration.

These cables also work in another way to alter cell structure. Each time a cell divides, the connection between the daughter cells becomes highly constricted until the two detach from each other. Multiple antiparallel cables of actin are laid down at the contact between the two daughter cells. The motor proteins in this case are linked to each other but are moving in opposite directions. The movement of motors that are connected to different actin cables will slide the cables in opposite directions. This movement generates constriction, like a purse string that is being pulled, and as it grows tighter, the opening shrinks until it closes completely.

External physical forces can also be exerted on cells and tissues. When soap bubbles are generated as a cluster, they reduce surface tension by organizing themselves in a structure that minimizes the area of contact among them. Like soap bubbles, cell membranes are composed of charged lipids. When cells assemble to form a discrete structure, the arrangement of their membranes can dictate their final pattern. When the arrangement of cells that give rise to the lens of the fly eye was documented, it turned out to be identical to the organization of adjoining soap bubbles, indicating that the physical forces that guide the arrangement of the lipid membranes also determine the final position of these cells (fig. 30).

It is interesting to consider other instances of physical rules that shape cells and tissues, where regulation of pattern operates merely by tweaking these physical forces. In Chapter 8 (see fig. 25) we saw how the identity of the germline of the worm is determined by concentration of structures termed P granules, which are not encapsulated by membrane, into the cell that will give rise to the germline. It turns out that this process is not executed by migration of existing granules from one side of the embryo to the other. Rather, these granules are continuously maintained by a dynamic balance between proteins that join them and proteins that leave them. This fine balance is guided by purely chemical and physical rules.

The asymmetric distribution of these structures is driven by a coarse gradient of a critical regulatory protein along the head-tail axis of the embryo.

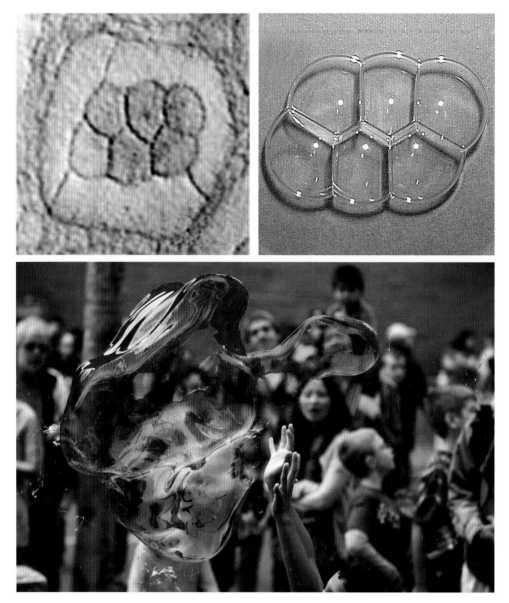

Figure 30. Physical forces contribute to shape the organism
The final shape of many organs is also dictated by physical forces.

Top: (left) A developing ommatidium in the fruit fly eye. Contacts between the membranes of the lens cells are minimized to reduce surface tension; (right) The pattern formed is identical to that generated by the equivalent number of soap bubbles (R. Carthew, Northwestern University); *bottom:* Popping a soap bubble (B. Shilo, Harvard Square, Cambridge)

This protein attenuates the capacity to form P granules in the head region of the embryo. As a result, these structures disassemble in the head region, and release their contents. The released components now follow a simple process of random diffusion and become trapped in P granules in the tail region of the embryo. This process eventually gives rise to the clear segregation of granules to that part of the embryo.

Contacts Among Cells

The process of embryogenesis is dynamic and amazing, as anyone who has followed it in live videos can attest. Some cells form in one region and migrate to another area. Others move relative to one another to change the organization of the tissue. All these movements resemble ocean fish schooling or a flock of gulls bobbing on a stormy sea, moving but maintaining the relative order of the group. The need to establish order and pattern in a continuously shifting landscape poses a complicated challenge to the developing embryo.

One strategy to maintain order is to keep cells of the same type tightly attached in the face of the surrounding movement and commotion. To stick together, the cell surface displays molecules that protrude externally and have the important feature of attaching to molecules of the same type presented by neighboring cells but not to molecules of a different type displayed by another group of cells. Thus, as part of its identity, the cell exhibits molecules that make it easy to associate with cells of the same type. This feature keeps cells together and, even after they mixed, allows them to regroup into a discrete and confined arrangement. One can think of the separate groups of cells as teams of football players who are distinctly marked by their uniform. Each team will gather with its coach after completing a play on the field (fig. 31).

The tight association among cells of the same type not only keeps them together but generates sharp boundaries between them and other cell types. Think of a family holding hands in a large and shifting crowd. Although the pressure of the moving crowd will promote separation between the family members, the family will strive to remain in the most compact organization to relieve the pressure and tension on the extended hands that are joined. Similarly, the cells of the same tissue attempt to reduce the strain on the con-

tacts between them and will therefore minimize the contacts with other cell types by aligning along a sharp border (fig. 32).

Cell Polarity

Imagine a field of sunflowers whose heads are all turned in the same direction. This is a beautiful image not only because of the combination of yellow flowers, green stems, and blue sky but also because of the sense of order it portrays. Every flower in the field turns in the same direction to face a global cue, the sun. Many tissues display polarity, meaning that the cells are all oriented in the same direction. Examples include the hairs on our skin and the feathers on a bird. Polarity is also detected in more intricate structures that are not visible to the naked eye. The hairs of cells in the inner ear, for example, are responsible for sensing sound at different wavelengths.

Generating a uniform polarity within a tissue is different from the allocation of distinct fates. Here the cells are coordinating, throughout the entire organ, either their growth or the construction of structures they produce, in order to create a uniform repeating shape. This global coordination of

◀ Figure 31. Generating compartments

As the organism develops, it is essential to generate restricted compartments that hold cells of one type together and prevent mixing with other cell types. Within the same compartment, cells display on their surface proteins that stick to similar molecules expressed by the neighboring cells. Whereas cells freely mix with cells of the same type, they minimize contact with cells from other compartments, creating sharp borders.

Top: Cells within the posterior compartment of every segment of the fruit fly (green) do not mix with their neighbors in the front compartment (C. Dahmann, Technische Universität, Dresden); bottom: Two groups in a football game segregate, maintaining cohesive interaction within the group and sharp boundaries with the other group (B. Shilo, Gloucester, Massachusetts)

Figure 32. Cells of the same type stick together ▶

Top: Cells within the left compartment (red), in the developing wing of the fruit fly larva, freely mix with their neighbors. However, they minimize contacts with the gray cells at the right compartment, forming a sharp straight boundary (C. Dahmann, Technische Universität, Dresden); bottom: Nuns belonging to the same order cluster at the ninth station in the Via Dolorosa (B. Shilo, Old City of Jerusalem)

Figure 33. Defining the global orientation
Emerging from a subway station in Manhattan, the passenger knows the coordinates of her location but not the orientation. Identifying a global cue, such as the position of the sun or the direction of traffic in the adjacent avenue, will provide her with the orientation (B. Shilo, New York City)

polarity may include cells that have assumed different fates and constitute distinct parts of the organ. In this respect, the cells must obtain information not on their precise position within the organ but on the global external co-ordinates. Imagine a woman coming out of the subway in New York. She will know the specific position of the street and avenue names. But as she emerges from the station, she will not know the global coordinates, such as which way north is and how to orient herself universally in her local position (fig. 33).

In the example of the sunflower field, the position of the sun provides the external signal regarding global directions and is very clear and visible to each flower. Thus, even though the individual plants aren't communicating among themselves, the whole field can behave in a highly coordinated manner. The generation of polarity within an organ is different in the sense that the signals regarding the global coordinates are weak, more akin to a hiker looking at the sky and trying to discern the position of the sun on an overcast day. Relying only on the external signal would therefore not induce a uniform response in all cells. To achieve a coordinated response, the cells communicate and reinforce their responses.

Imagine a large crowd of people standing in an airfield, waiting for a plane to land. Before the plane is visible, the viewers scan the sky in different directions and try to hear the distant noise of the engine. Once the first person or group of people see the plane, the other viewers orient their gaze according to the direction where the fingers are now pointed, even if they do not yet see the plane themselves. Eventually, the entire crowd will be facing the same direction.

Again, the fruit fly comes to the rescue to decipher the biological mechanisms that regulate cell polarity. The wing of the fly is composed of two cell layers that adhere to each other. Each cell extends a single hairlike structure, and amazingly, all hairs are pointed in the same direction. The orientation of these hairs may be important for aerodynamic streaming along the wing during flight (fig. 34). It is thus easy to detect when these hairs are disoriented, which is comparable to the difference between a person whose hair is tightly combed versus one whose hair is disheveled. This assay allowed identification of mutations in genes that are important for coordinating and maintaining the polarity of the cells.

The principles that have emerged from these studies reiterate the impor-

Figure 34. Defining a global orientation of cells ▶
Global coordination of cell orientation in a tissue may be achieved by a response of every cell to a general signal that will "point the way." When that signal is prominent, coordinated orientation can be readily achieved. But if the global signal is weak, cells need to communicate with one another to enhance and refine the signal so that they can all orient themselves in the same direction.

Top: Developing hairs on the wing of a fruit fly. The boundaries of each cell are defined by hexagonal arrangement of the membranes, and every cell generates a single hair. So that all the hairs will point in the same direction, the cells respond to a weak global cue and communicate to enhance it (D. Ma and J. Axelrod, Stanford University); *bottom:* A field of sunflowers all pointing in the same direction in response to the orientation of the sun. The strength of the external signal alleviates the need for the plants to coordinate among themselves to generate a uniform polarity pattern throughout the field (B. Shilo, Lepakshi, Andhra Pradesh, India)

CHAPTER 11

Stem Cells

The intricate processes of cell and tissue differentiation are used even after embryogenesis ends. Only a few tissues, such as the lens cells of the eye, are maintained without replacement throughout the lifetime of the organism. Many tissues in the adult organism are continuously renewed. The most obvious examples are tissues in which we are blatantly aware of continuous growth. Barbers and hairdressers make their livelihood because cells at the base of each hair divide, differentiate, and die, leading to the growth of hairs. Most cells in our circulatory system have short lifetimes; for example, the lifetime of a red blood cell is 120 days. Our skin continuously renews itself by shedding dead cells from the surface after new differentiated cells arrive. The cell layer defining the intestinal lining is completely renewed every couple of days. In the case of the skin and intestine, this renewal is essential because of the harsh insults faced by these cells. These tissues are hard at work maintaining their cell composition and integrity over the lifetime of the adult organism.

The actual cells that will replace the differentiated cells are not generated in advance during embryogenesis. Simply put, each organism over its lifetime needs too many cells for them to be generated in advance. So we carry with us, not the actual "replacement parts," but only the capacity to generate them when needed. Stem cells represent this potential to generate the differentiated cells when needed and ultimately develop different kinds of cells, tissues, and organs. Two basic features define stem cells. On the one hand, they can remain in a nondifferentiated, proliferative state. On the other hand, they have the ability to generate differentiated cells.

The Niche

How do stem cells fulfill these two opposing requirements? In fact, these conflicting features cannot be maintained within a single cell and are segregated into two sibling cells. The prototypic stem cell is proliferative and nondifferentiated. On division, the two daughter cells become distinct. One cell will remain proliferative and nondifferentiated, just like the parent cell, and the sibling cell will go on to differentiate. Maintaining this fine balance between the two types is critical. Excessive differentiation will deplete the proliferative cell pool that embodies the potential for future differentiated cells. Conversely, skewing the balance toward the nondifferentiated cells will give rise to the depletion of the differentiated cells, and possibly to a tumor because of excessive proliferation. Identifying and understanding the mechanisms that keep this fine balance have been the focus of extensive research.

The basic idea is that the process of maintaining the stem cells in their nondifferentiated form requires continuous input from their environment (that is, from other cells). The stem cells must be continuously exposed to signals that will keep them in that state. This specific environment with the signals it provides is termed the stem cell niche. In some instances, it may represent a physical entity that is morphologically distinguishable. In other cases, the niche may simply result from distinct neighboring cells that produce the signals to maintain "stemness." The stem cells can sense these signals only in the immediate environment of the cells that produce them. After a stem cell divides, there is physical room for just one cell in the niche, and the other cell is excluded. Once the excluded cell is no longer exposed to the signals that maintain it in the stem cell state, it will go on to differentiate. This mechanism controls the environment of the stem cells and influences the asymmetric features of the stem cell progeny (fig. 36).

The niche concept may be compared to inheritance systems in many agricultural societies in which the father cannot divide his land among his heirs. Only one son inherits and remains on the land, which for our purpose represents the niche. The others are forced to leave the land and "differentiate" into other professions that will provide their livelihood.

Elaborate mechanisms have evolved to maintain the niche as a tight entity. In some cases, one of the daughter cells physically sticks to the niche while the other is pushed away. In other cases, special mechanisms restrict the diffusion of the signal emitted by the niche cells, so that only the cells immediately nearby will encounter it. Because of the two environments, each daughter cell will receive distinct instructions that guide it in different directions.

The beauty of the niche system is that it regulates the number of stem cells that will remain, regardless of how many divisions they undergo. In other words, once the niche environment is properly maintained, the progeny of the dividing stem cells are simply responding to this fixed environment to maintain the ratio between the two cell types.

Another attractive feature of the niche system is that the global consequences are not directly linked to each individual stem cell but can also be seen at the population level. Suppose that the two progeny of a stem cell exit the niche and differentiate. This departure will leave a free spot in the niche that could be taken by the daughter of a different stem cell, but the total number of stem cells will still be maintained. Flexibility at the population level can thus compensate for glitches that may occur at the level of individual stem cells, thus maintaining a fixed global ratio of stem cells versus differentiated cells.

Cells that continuously proliferate present another potential problem.

Figure 36. *Stem cells and their niche* ▶
When a stem cell divides, one daughter maintains the stem cell fate while the other produces a differentiated progeny. Stem cells are positioned in a restricted spatial niche that provides signals maintaining them in a proliferative, nondifferentiated state. After division, only one progeny is retained in the niche and the other differentiates.

Top: Once the eye of the zebrafish (*Danio rerio*) is specified, cells differentiate and produce neurons that sense light. Retinal stem cells (red) are maintained throughout the animal's life in a niche located next to the lens (K. Cerveny and S. Wilson, University College London); bottom: The live yeast stock, or mother, is carefully maintained in the bakery. Portions are allocated to produce dough and bread. A fine balance is kept to ensure a steady production of bread while continuing to propagate the yeast stock (B. Shilo, Hi Rise Bakery, Cambridge, Massachusetts)

Despite the high fidelity of DNA replication, mistakes do occasionally occur. A cell that undergoes many divisions throughout its history may thus accumulate deleterious mutations. This problem is even more serious in the case of a stem cell, because it continuously produces new progeny that will carry the same mutations. One solution for this hazard is based on the mechanism described above—namely, stem cells can replace each other in the niche. The presence of sufficient continuous replacements diminishes the chance that a deleterious mutation will be fixed in the organism.

The study of stem cells has been complicated because in some tissues the niche is not morphologically distinct, and the number of stem cells may be small. In addition, true stem cells are identified by their behavior rather than their morphology. The best operational test for stem cells is that, when genetically marked, they will continue to give rise to differentiated cells over the long term. A marked cell that is not a stem cell will differentiate and, in the absence of renewal capacity, will eventually disappear from the body.

The general concept of stem cells and their niche holds for very diverse systems in a wide range of organisms. It underlies the continuous production of sperm cells in the testes of vertebrate and invertebrate organisms and the ongoing cell renewal in the circulatory system, skin, and intestines. But although the concept of the niche applies to different biological systems, each of these tissues and organs have unique characteristics. The distinctions include the physical structure of the niche that relies on the architecture of each tissue. In addition, there are differences in the number of stem cell divisions and their trigger, which may be continuous (such as production of sperm cells) or may arise only after an insult to the tissue, thus being controlled and induced by the cells that are already differentiated. For example, muscles harbor dormant stem cells that divide and rebuild muscle mass after injury.

The analysis of stem cells from a variety of tissues and organisms has provided a broader understanding of their behavior and regulation. This is an essential basis for applying the knowledge to therapeutic uses. Such practices are already prevalent for the circulatory system. Through a stem cell transplant from a healthy donor, a patient's white and red blood cells can be permanently replaced and restored. Millions of people who have had bone marrow disorders and cancers such as leukemia owe their lives to this procedure. The challenge is to be able to apply a similar methodology to reconstitute other organs. This involves first identifying the stem cells for a given tissue and then understanding how to direct them to differentiate so that they will produce a replacement for the missing or defective organ.

Type 1 diabetes is a disease prevalent among young people. It involves an irreversible destruction of the cells, known as beta cells, which produce insulin in the pancreas. For the rest of their lives, diabetics must depend on injections of insulin several times a day to keep functioning. Whether there are stem cells that can generate the missing beta cells is still debated. But clearly, if such cells could be identified and the method to trigger them to differentiate into insulin-producing cells figured out, it could provide a long-awaited solution for millions of people.

"Reeducating" Differentiated Cells

All our discussions about embryonic development so far have had a clear time arrow, in which cells always progress forward toward more defined and specific fates. The cell formed on fertilization of the egg divides numerous times, and the daughter cells encounter multiple junctions at which developmental decisions are made. At each junction the new cells acquire greater specificity and distinction, which is reflected in the array of genes and proteins they express. But despite this specification, each cell maintains the original genetic information of the fertilized egg, which embodies the "instructions" for making a complete embryo. Since this information is retained, the question that comes to mind is, Can cells actually "go backward" to a less differentiated state?

This fundamental question addresses the ability of cells to change their pattern of gene expression and return to a less differentiated state. Is this at all possible? And if so, do the cells have to progressively trace their steps backward in reverse order, like Hansel and Gretel in the Brothers Grimm fairy tale, who followed the pebbles they had dropped on the way into the forest in order to return home? Alternatively, can cells simply erase at once all information regarding their current state of differentiation, generate a clean slate, and start anew?

When addressing this issue, it is important to note that genetic material is altered in several layers. The most basic level is the sequence of bases in the

DNA, which is retained in DNA replication and faithfully transferred through the sperm and egg to the progeny. Mutations in the DNA can lead to deleterious consequences, including such diseases as cancer. In evolutionary terms, DNA alterations that modify and improve a particular trait in the context of a given environment drive the diversification and speciation of organisms. For our discussion we can regard this level of information as fixed and unchanging within the organism.

Another layer is the cells' use of the genetic information in diverse and dynamic ways. As we have seen, as they differentiate cells express distinct classes of genes as RNAs and proteins based on their developmental fate and tissue context. The process of cell differentiation is versatile and can be readily altered. Think of an airplane factory, in which the master plan containing all the information is stored in the central computers. Different parts of this plan can be copied and distributed to computers in the relevant production departments, depending on the demand at a given time.

A third tier of DNA regulation, imposed between these two layers, is termed epigenetic modification. This tier, though not part of the actual code of bases within the DNA, increases the stability and sustainability of the cell fate. In a sense, it allows the cell to fix a certain pattern of gene expression. If we think of the airplane factory again, this is comparable to marking highlights on particular pages of the plan files. These markings do not change the information in the plan but tag these pages for preferred selective usage.

How is this epigenetic modification manifested? Stable modifications of DNA can take place without altering the base composition. One prominent example employs DNA methylation, a biochemical process in which a small methyl group (consisting of one carbon and three hydrogen atoms) is added to defined bases within the DNA. The physical modification of the DNA adjacent to a given gene reduces the propensity of this gene to be copied to RNA and thus affects the level of its expression within the cell.

Although we typically picture DNA as two strands forming a double helix, it is actually stored much more compactly in the nucleus of the cell. We read earlier that each cell in our body contains a DNA strand that, when stretched, will reach up to one meter in length. When we consider that the cell nucleus is a million times smaller than that, it is evident that there is a packaging challenge. As an analogy, consider a thousand-volume encyclope-

Figure 37. Epigenetic marks are difficult to erase
In addition to the information embedded in the genetic material inherited by the cells, external modifications are added to the DNA during development. These marks define areas on the chromosomes that are more or less prone to be expressed according to the fate of each tissue. To drive a differentiated tissue back to the nondifferentiated pluripotent state, the marks that were introduced on the DNA must be removed, generating a "clean slate." The removal of such marks, as is the case for tattoos, is not a trivial task (B. Shilo, *28 Seeds* musical, Boston)

dia containing the same amount of information as that of the DNA within the cell. If all pages of the encyclopedia were taped to one another and laid out in sequence, the resulting trail of paper would extend more than 200 kilometers. Clearly, a more compact storage solution is to bind the pages into a thousand volumes and put them in a single library room.

The DNA is covered with proteins called histones that regulate its folding and compaction. These proteins can also be modified in specific ways. Proteins that are attached to distinct parts of the genome can specifically tag these regions for being expressed or being silenced. Retaining this information through every cycle of DNA replication keeps this tagging fairly stable.

Returning to our question of whether differentiated cells could "go back" to their primordial, undetermined state, we have to ask ourselves not only whether the actual pattern of gene expression can be reversed but also whether the various epigenetic modifications on the DNA of a differentiated cell, discussed above, can be erased, and if so, in what order (fig. 37). Considering our analogy of the master plan for producing an airplane, can the highlighting marks be removed so that the plan can be used to produce copies of other parts of the airplane?

In 1962, Sir John Gurdon published a critical experiment showing that a reversal of differentiation is indeed possible. In this work, Gurdon transferred the nucleus of a differentiated cell from a frog tadpole intestine to an unfertilized egg that was depleted of its original genetic material. The striking result was the formation of a complete tadpole. We refer to this experiment as cloning, during which the nucleus of a single differentiated cell in the body was used as the basis for forming a complete and identical organism. This experiment reaffirmed the notion that all cells in the body have the information necessary to produce an entire organism. In addition, it proved that cells can revert to an original, nondifferentiated state that would allow a single cell to give rise to a whole organism, just like the fertilized egg.

A similar experiment, now in a mammal, was performed in 1996, and gave rise to the world's most famous sheep, Dolly. This time a nucleus from a differentiated mammary gland cell was transplanted into an oocyte that was depleted of its DNA. This cell was transplanted in a foster ewe mother who supported embryonic development and gave birth to Dolly. Dolly subsequently lived for six and a half years before dying of lung disease. She excited the world's imagination with the ability to clone a mammal, as well as the science fiction–like possibilities it opened for one day cloning humans. For scientists, the major revelation was that the differentiated cells of a mammal can revert to become pluripotent.

The experiment with Dolly also addressed unease regarding the use of a mature cell to form a new embryo. Because cells divide numerous times, even though the information encoded by the DNA is replicated faithfully, there was concern that elements that maintain the structure of the chromosome, such as sequences that cap the ends of the chromosome, could become modified. If this happened, then mature chromosomes would not be able to faithfully undergo the number of divisions required to produce the embryo.

As an analogy, think of Albrecht Dürer's famous *Rhinoceros* woodcut print from 1515. The woodcut was used to make numerous prints, and later prints suffered in their quality because they showed cracks and imperfections that had accumulated in the wood template.

The generation of a complete and healthy sheep from a differentiated nucleus showed that chromosomes from a mature cell can still undergo the number of divisions necessary to make an embryo, and remain intact. The process of cloning (in sheep and subsequently in other mammals) succeeds only on rare occasions, however. One reason for this low efficiency of success may be that the chromosomes in many of the nuclei tested have not been capable of facilitating the enormous number of divisions necessary to make a new embryo.

Nuclear transplantation experiments are remarkable in demonstrating that it is possible, albeit at a low rate of success, to drive nuclei back to their original nondifferentiated state and then carry out the ultimate task of producing an intact organism. But for years these experiments remained highly challenging, since they involved a drastic experimental manipulation whose consequence was monitored only after many subsequent steps that took place, by the ability to produce a whole embryo. In other words, there was little control or knowledge about the actual biological processes that push the nucleus back to the nondifferentiated state.

In 2006, however, the lab of Shinya Yamanaka in Japan carried out a revolutionary experiment in which defined genes that could drive differentiated cells backward were identified from a large battery of genes tested. For this breakthrough, he was awarded the Nobel Prize in 2012, together with John Gurdon. Yamanaka found that the expression of four defined genes, all of which encode proteins that affect gene expression, could drive differentiated cells back to their naive state. This includes not only the pattern of genes expressed by the nondifferentiated cells but also the complete erasure of epigenetic marks on the chromosomes that are typical of differentiated cells. The four genes that are required reinforce one another's expression and thus stably maintain the nondifferentiated state. These nondifferentiated cells, termed iPS, short for induced pluripotent stem cells, are now capable of generating an entire embryo (fig. 38).

The implications of this experiment are enormous. They point to a defined mechanistic circuitry that drives cells back to their nondifferentiated state. This now allows us to explore, at the molecular level, the essence of maintaining cells in a nondifferentiated state. In addition, it provides the potential to generate such a pool of cells at will, rather than relying on the restricted supply of stem cells harbored in different tissues of the body. The consequence is that cell origin does not matter. As long as the cell can be driven back to its nondifferentiated state, it can then be directed to form the organ of choice. This could lead to personalized medicine in the broadest sense, in which it would be possible to generate "spare parts" from one's own body, overcoming not only the difficulty of obtaining organs for transplantation but the immunological barriers of rejecting organs from a foreign donor.

In many respects, this personalized medicine is already being practiced for tissues such as the circulatory system. Doctors can draw peripheral blood containing differentiated cell types from a patient. Then, hormones are added to drive the cells to a less-differentiated state that can reconstitute all the cell types in the blood system. Such manipulations are saving the lives of patients every day. They are successful because we have accumulated throughout the years a detailed knowledge of the molecular pathways that control the differentiation of blood cells. In addition, it is easier to manipulate a tissue, such as the blood system, that is organized as a suspension of cells. Dealing with tissues that have a distinct three-dimensional architecture and consist of multiple layers will require more elaborate knowledge and manipulation. But we should certainly expect to see dramatic advancements in our lifetime.

Figure 38. Induced pluripotent stem cells (iPSs)   ▶
Fully differentiated human cells can now be driven back to their nondifferentiated stem-cell state and then be induced to form a different cell type. The development of human iPS systems in recent years has opened a vast new field for exploration, thanks largely to the remarkable ability of iPSs to generate an unlimited source of any human cell type.

Top: Human iPSs derived from skin were used to generate neural stem cells that form cellular rosettes with the dividing stem cells (pink) at the center. These structures mimic normal differentiation of neuronal cells and enable studies on the formation of the human forebrain (Y. Shi and R. Livesey, Gurdon Institute, Cambridge, England); *bottom:* A group of people who lost their jobs train for a new profession (B. Shilo, Artisan's Asylum, Somerville, Massachusetts)

Natural Regeneration

When we consider the promise of creating human organs to replace defective parts, we can look to organisms that are successful in regenerating organs after an injury. The ability of certain animals to replace missing parts is a dramatic, yet poorly understood aspect of biology that has intrigued scientists since the 1700s.

Salamanders, frogs, and some fish have the amazing ability to regrow substantial parts of the body. The amputation of appendages results in the restoration of the correct cell types and form. The salamander, for example, faithfully regenerates the missing limb segments amputated anywhere along the limb axis. This accurate regeneration is no mean feat, considering the complexity of the limb and the diversity of tissues that give rise to the different elements. During embryonic development, the skin and nerves originate from the outer cell layer, while muscle, bone, and blood vessels are derived from a more internal cell layer. These animals are also capable of restoring damage to internal organs such as the heart or kidney, although the exact organ form is not always reconstructed.

More primitive animals show an even broader capacity to regenerate. Hydra are cnidarians that live as freshwater polyps. They have two cell layers and a simple pattern composed of the tentacles, a mouth opening, and a foot. Even tiny body fragments of the hydra are capable of regenerating the polyps. Moreover, when the hydra cells are dissociated, they can reaggregate and produce a new polyp.

The champion of regeneration is the flatworm known as the planarian. This flatworm is able to regrow a new head, tail, sides—even a whole organism—from tiny body fragments, as biologists showed at the end of the nineteenth century. In fact, Thomas H. Morgan, the founder of fruit fly genetics, carried out experiments on the regeneration of planaria more than ten years before he discovered the first mutation in the fruit fly. This regeneration capacity is striking in view of the complexity of the flatworm, which possesses an intricate anatomy, including a nervous system, muscles, epidermis, eyes, and intestine. Pieces as small as 1 percent of the total body (constituting approximately ten thousand cells) are able to regenerate an entire planarian within a week (fig. 39).

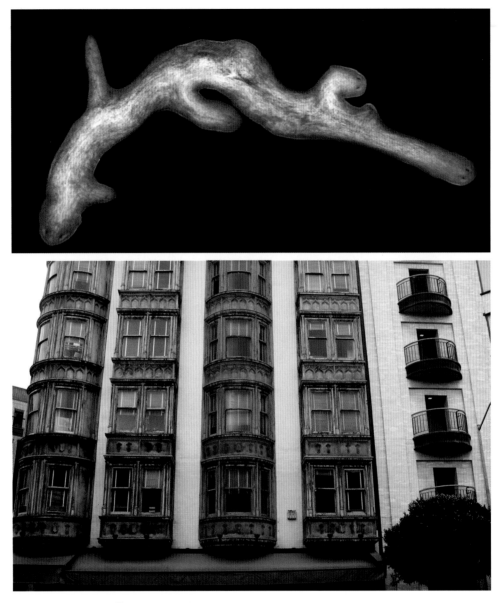

Figure 39. Tissue regeneration

Some organisms maintain the ability to regenerate complete organs following injury through the use of their stem cells. The most striking example is the flatworm, or planarian, which can regenerate its entire body plan from any small body fragment.

Top: Incisions along the planarian body led to the generation of a six-headed worm (P. Reddien, Whitehead Institute MIT and Howard Hughes Medical Institute); *bottom:* Using the old patterns to shape a new addition to a building on Kearny Street in San Francisco (B. Shilo)

These striking phenomena raise a host of questions. First, how does the tissue "sense" the damage and trigger recruitment of the elaborate regeneration process accordingly? Second, what is the mechanistic basis for tissue regeneration, and how are spatial cues for the formation of new tissue provided? Last, do the different organisms employ the same mechanism, and are there distinct strategies for repairing different organs? Because little is currently known about the damage-sensing mechanisms, I focus here on the deployment of regeneration programs.

The most plausible way to consider organ regeneration is through the recruitment of stem cells. Because these cells are nondifferentiated and have the capacity for self-renewal, when they receive an injury trigger they can replicate and produce the required differentiated progeny. The most critical question in understanding regeneration is the nature of these stem cells. If kept in the most primordial state, a single stem cell may be capable of giving rise to a diverse set of differentiated cells that would consist of the entire organ or even the complete animal. Alternatively, if the stem cells are already specified as a given cell type but still retain their proliferative capacity, then they would be able to contribute only a limited repertoire of cell types to the regenerating organ. It follows that, to produce an entire organ, different types of stem cells would need to cooperate.

The burning experimental issue in current research on regeneration is to identify the origin of the cells that build the new organ. Are they descendants of a single stem cell, or do they come from different origins? Planaria are amenable to explore these questions experimentally. It is possible to irradiate one animal, thus blocking the ability of its cells to divide and contribute to regeneration. Cells can then be transplanted from a normal planarian to the irradiated host and examined for their regeneration potential. When the technique is pushed to its limit, even a single cell can be transplanted and assayed for its regenerative capacity within an irradiated host flatworm. The emerging results indicate that single transplanted cells are sufficient to lead to regeneration, implying that they have a very broad capacity and are normally maintained in a primordial state. Such cells circulate within the body of the planarian, and on injury they can divide and generate a wide repertoire of cell types that restore the missing organ or even a whole organism.

Does the same principle hold for regeneration in more complex animals such as salamanders and frogs? In these animals, a scar tissue called the blastema is formed at the site of amputation. This tissue provides the source of cells for the new organ. A variety of experiments demonstrate that the blastema contains more than one type of regenerative cells. In fact, each tissue layer regenerates separately. Distinct populations of stem cells thus contribute the different cell types to the blastema. To reproduce the organ faithfully, tight coordination between the different kinds of regenerating cells is essential.

In conclusion, in simpler animals the concept of "one stem cell does it all" applies. Conversely, in more complex organisms distinct populations of stem cells, each having a more restricted differentiation program, contribute to the formation of the complete organ. This distinction is clearly linked to the complexity of the organism. As an analogy, a single plumber can repair a faulty kitchen sink, but renovating a whole building requires teams of specialists and tight coordination among them.

Future research will likely involve the improved definition and marking of the stem cells in each system, as well as a detailed mechanistic understanding of the regeneration process. It will examine whether the two regeneration strategies are mechanistically distinct or whether they represent variations on a theme. We would also like to know which spatial coordinates are used by the regenerating tissue in order to produce an elaborate new structure, in an organism that is already fully developed. Last, identification of the signals that alert the cells to tissue injury and initiate the regeneration process is eagerly awaited. A deeper understanding of these recruitment signals may provide more precise and efficient ways to trigger the formation of new organs in culture.

CHAPTER 12

What's Next?

Tremendous progress has been made in the field of developmental biology since Spemann and Mangold's experiment in 1924, which laid the foundations for a deep understanding of embryonic development. At the conceptual level, scientists have unraveled the communication pathways among cells and proved that all multicellular organisms share these pathways. In addition, stem cells have been identified for a variety of tissues and now allow us to reconstruct these tissues artificially.

At the research level, we can conduct experiments today that we could not even dream of one or two decades ago. Scientists have fully sequenced the genomes of many multicellular organisms, allowing us to know and compare the information for the structure of all proteins made by each organism. We have determined and compiled the expression pattern of all genes in a variety of tissues during development, which enables us to associate groups of genes with defined biological processes and understand why they are coordinately expressed. We can eliminate gene function at will in the whole organism or in a defined tissue or restricted time window in a range of organisms, including such higher vertebrates as mice. This allows us to assess the role of each gene critically by inferring what goes wrong in its absence.

Using sophisticated microscopy and imaging tools, we can monitor the behavior of live cells over time inside a developing embryo, as well as the dynamic behavior of proteins within cells and organisms. And we can create stem cells from differentiated cells that were already recruited to form another organ in order to produce an unlimited supply for use in the artificial generation of organs. These advances and achievements are truly remarkable. But what can we expect in the near and more distant future?

Scientists today hope that for each tissue we will ultimately possess knowledge and understanding that are deep enough to allow us to control each tissue and direct its growth. If we can identify and isolate the stem cells, which represent the primitive, nondifferentiated state, we will be able to direct the differentiation of these cells to form any organ needed. This will be a tremendous leap for medicine, because a patient's cells could be used to generate organs anew and render irrelevant the immunological barriers that represent the current obstacles to organ transplantation. Success in this promising avenue would depend not only on identifying the stem cells but also on achieving the ability to control their differentiation in a reliable manner.

The ultimate goal for tissue engineering is to generate new organs from the cells of an individual patient. The challenge becomes even more formidable for tissues that have a complex architecture involving specialized cell types. Think, for example, of the cells that secrete insulin in the pancreas and are concentrated in clusters or "balls." These clusters, termed the islets of Langerhans, contain five different types of cells that secrete proteins into the blood; one of them secretes insulin (fig. 40). Today, scientists are working to restore these islets in patients with type 1 diabetes, who have lost their insulin-producing cells.

Eventually, we will also be able to control the state of the cells in the opposite direction — namely, to take differentiated cells and have them revert back to their original nondifferentiated state. With the development of such a procedure, we would no longer depend on a limited source of stem cells and would be able to generate an unlimited supply of nondifferentiated cells. To carry out this procedure reliably, we need to understand better how cells dedifferentiate so that we can avoid such harmful consequences as the uncontrolled growth or premature aging of the cells.

Another aspiration is personalized medicine tailored and adapted to the individual genetic makeup of each patient. Even today, the rapid advances in DNA sequencing technologies allow us to obtain a complete DNA sequence for each person at a reasonable cost. And these costs will continue to plummet in the coming years. The DNAs of individual people differ by only about

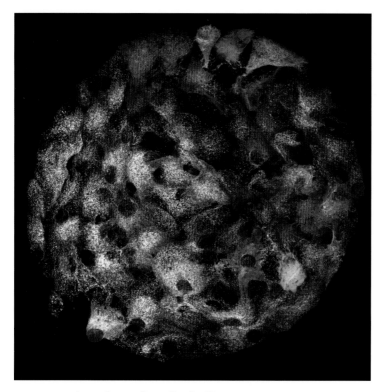

Figure 40. Insulin-producing cells: Will we be able to imitate nature?
The complexity of human tissues is viewed with a different perspective when we consider the challenge of rebuilding them in the lab, starting from the nondifferentiated cells of a patient. The image shows a single islet of Langerhans from a mouse pancreas. Five different cell types (including the beta cells that secrete insulin) contribute to form the final structure. Blood vessels penetrate the structure and carry secreted insulin to the body (E. Geron and B. Shilo, Weizmann Institute)

one in a thousand bases. Some of these differences are critical and define the characteristics that distinguish us from one another. They relate to health and disease, and define our susceptibility to different drugs, resistance to other medications, the probability of encountering certain diseases, and more. As we learn more by correlating diseases to their DNA sequences, it will become possible to predict how our bodies will react from looking directly at our DNA sequence.

Many aspects of medicine will remain empirical, despite this wealth of information. For example, it will still be difficult to know with certainty which drug or combination of drugs will elicit the best response in a given patient's cells. When it comes to bacterial infections, we can grow the infecting bacteria in a culture, test which antibiotic combination will be most effective, and then give the same combination to the patient. But the challenge becomes more complex when we apply similar approaches to treating a defective organ inside a person.

This is where stem cells again come to the rescue, not to create new organs for replacement, but to generate an organ in culture that will provide a testing ground for the best strategy to treat the diseased organ of the patient. In this case, only few restrictions will apply, since the organ that is artificially formed will never be transplanted into the patient. These artificial tissues can be compared to the crash test dummies used to examine the effectiveness of safety devices that protect passengers in car accidents.

The possibilities are endless. For example, when treating a patient with Crohn's disease, cells from the patient would be used to generate an artificial bowel, and a battery of treatments could be tried on this test organ. Similarly, nerve tissues could be formed to examine treatments for neurological disorders. The main breakthrough is that the treatment will be tailored for the cells and organs that come from the patients themselves. As a result, the conclusions reached will represent the best possible treatment for the individual patient, as opposed to an average best treatment.

Despite the detailed knowledge we have today regarding the molecules and mechanisms that guide development, many basic concepts remain a mystery. These unknowns will fuel the next wave of research and provide the impetus for new discoveries and advances. I outline some of these open questions here.

The genetic code for translating an RNA sequence into proteins is clear and unambiguous. This is why we can be certain which protein the RNA will form for every coding region in the DNA. The situation becomes much more complicated when we consider the sequences on the DNA that regulate the organization and expression of genes. Although we know the full sequence of genomes for different organisms and the pattern of genes they express, we can compare this knowledge to recognizing the individual words in a sentence without understanding the grammatical structure or the deeper mean-

Glossary of Scientific Terms

actin. A protein monomer that is assembled in a polymeric structure that forms linear or branched microfilaments. The microfilaments mediate such cardinal processes as cell motility, cell shape, and cell division.

cell communication pathway. A set of proteins that mediates the transmission of signals among cells. Typically one cell sends the signal in the form of an activating protein, which is recognized by a receptor on the membrane of the receiving cell(s). Additional proteins transfer the information from the membrane to the cell nucleus, where they usually alter the pattern of gene expression.

cell context. The state of the cell during a distinct phase of development or within a particular tissue. The composition of the cell affects how it interprets external signals and the types of genes that will be expressed.

cell lineage. The developmental history of a tissue or part of an organism from particular cells in the embryo through to their fully differentiated state.

cell migration. Movement of cells from one position to another.

chromosomes. A condensed organized structure of DNA and proteins that lies within the cell nucleus and harbors the genetic material.

collective cell migration. Coordinated movement of a group of cells as a sheet or a cluster.

cytoskeleton. The cellular scaffolding of protein filaments and tubules within the cytoplasm of the cell. This microscopic network is responsible for cell shape, movement of components within the cytoplasm, cell motility, and division.

default developmental path. A developmental program that cells will follow in the absence of additional external signals from neighboring cells.

DNA. Short for deoxyribonucleic acid, this is the molecule comprising the genetic material in all cells and organisms. It is organized as a double helix and is capable of providing information for the production of RNA and proteins, in addition to replicating itself.

Drosophila melanogaster. A species of the fruit fly that has been used as a model genetic organism for the past century. The species name is Greek for "dark-bellied dew lover."

embryogenesis. The intricate process by which a fertilized egg forms a multicellular embryo with its elaborate array of tissues and organs.

epigenetic modification. Semistable alteration of the genetic material that is superimposed on the basic sequence of nucleotides in the DNA. This process is executed primarily by proteins that are bound to distinct regions on the DNA.

evo-devo. Short for evolutionary developmental biology, an area of study that aims to understand the diversity of patterns among different organisms, and the mechanisms by which they arose.

evolution of pattern. Formation of divergent structures either within the same species but more commonly during the generation of new species.

execution of cell fate. Formation of a distinct cell type or organ by the expression of the full array of genes necessary for its structure and function.

feedback regulation. Modulating the rate or efficiency of a process, based on the accumulation or absence of the final product.

gene. A stretch of nucleotides on the DNA that is responsible for the production of an RNA molecule that will give rise to a protein. This is the main unit of heredity on the chromosome, and a human cell has a total of ~25,000 genes.

gene coding sequence. The part of the gene that provides the information for the sequence of amino acids that will be assembled to produce the protein.

gene expression. Formation of RNA from the gene; this RNA sequence will serve as the template for the production of protein.

gene redundancy. Genes that carry out overlapping functions so that one gene can "cover" for the loss or malfunction of the other. Typically, redundant genes generate proteins that are structurally similar, but redundancy can also be generated by distinct proteins that carry out overlapping functions.

gene regulatory sequence. The part of the gene that dictates when, how much, and in which cell type RNA will be formed, based on the DNA sequence of this region.

genetic screen. Analysis of a large collection of organisms of the same species, each

harboring a different mutation or mutations, to identify the mutations that will give rise to alterations in the tissue or process of interest.

genetics, forward and reverse. Forward genetics examines a large set of mutations for defects in a particular biological process to identify the genes that are critical. Reverse genetics starts from a known gene of interest and looks for the biological consequences of removing this particular gene.

genome. The complete genetic information of a cell.

genotype. The genetic makeup of the cell, which will dictate its shape, activity, and behavior.

hierarchical patterning. A temporal process in which cell fate choices are allocated progressively over time, such that the earlier instructions are coarser but affect a larger spectrum of cells.

homeobox. A defined domain within homeotic proteins that mediates their association with DNA. The protein sequence comprising the homeobox is conserved across all multicellular organisms and is critical for their capacity to induce gene expression.

homeotic transformation. Conversion of one body part to another part normally found in a different region of the organism.

homeotic genes. Genes responsible for conferring the distinct identity of different body parts.

hormone. A secreted protein that transmits information by binding a receptor protein embedded in the membrane of the cells receiving the signal.

induced pluripotent stem cell (iPS). A differentiated cell that was reverted to its nondifferentiated state, such that it can now be directed to form a different cell type or organ.

induction. Communication among cells in a developing embryo, which alters their gene-expression program and dictates the final structure of the embryo. This concept was originally identified when cells from the future head of a newt embryo, on transplantation to a recipient embryo, gave rise to the formation of a second head by recruiting cells of the recipient.

master regulatory gene. A gene that dictates the global fate of a tissue. The presence of the protein it produces provides the tissue "context" in which incoming signals are interpreted, such that they give rise to the expression of genes unique to that tissue.

microbiome. The totality of microbes, their genomes, and the environment they create in a particular niche such as the human intestine.

module. A repeated unit that can be used in diverse biological contexts. The concept of a module is broadly applied in biology, from the definition of distinct domains within a protein, such as the homeobox, to repeated structures within the whole organism, such as vertebrae or segments.

morphogen. A secreted protein that is distributed within a tissue in a decreasing concentration at a distance from the source. Different cell fates are induced based on the level of this protein, such that a single molecule can lead to the formation of several cell types within the tissue.

multicellular organism. An organism that contains many cells that adopt different structures and functions. Through their coordination and diversity, these cells contribute to the complexity of the entire organism.

mutation. A change in the sequence of the DNA. This change may be as subtle as a single nucleotide alteration or as severe as the loss of a long stretch of nucleotides. Depending upon its position and nature, the mutation can alter the expression pattern or the structure and function of the protein that is formed.

myosin. A protein that uses chemical energy to slide across actin cables. This process guides the motility of cells, changes in cell shape, and the transfer of components from one part of the cell to another.

nucleotides. The building blocks of DNA, represented by four types (A, C, G, and T). Their linear order determines the regulation of each gene and the structure of the resulting protein.

nucleus. The part of the cell where chromosomes are maintained and where the production of RNA takes place.

phenotype. The observable characteristics or traits of an organism, notably after alterations in the genetic material.

preformation theory. The notion that organisms develop from existing miniature versions of themselves. According to this theory, pattern formation involves only the "growth" of the different organs, rather than their actual generation.

primordial multicellular organism. The primitive organism that existed before evolutionary divergence of the vertebrate and invertebrate phyla. The notion is that many of the elements that are common to all multicellular organisms reflect components that were already present in the last common ancestor and were preserved in evolution owing to their central roles in all organisms.

protein. A biological molecule composed of building blocks termed amino acids. The linear sequence of the protein is dictated by the sequence of the gene. Reproducible folding of the protein chain, guided by its primary amino-acid sequence, will determine its three-dimensional structure and ultimate function.

receptor. A protein embedded in the cell membrane. The structure of the receptor is altered following binding to a protein that is externally displayed to the cell, triggering the activation of proteins within the cell that relay the information to the nucleus.

regeneration. The process of reforming an organ or tissue following an injury.

ribosome. A large complex of proteins and RNA molecules located in the cell cytoplasm. This structure uses the information encoded by RNA molecules to assemble and link the building blocks of proteins. A cell can contain up to ten million ribosomes.

RNA. Ribonucleic acid, composed of four nucleotide building blocks. The major function of RNA is to serve as an intermediary between the information encoded in the DNA and the production of proteins. RNAs also fulfill structural roles in the cell and provide components essential to the ribosome for translating the information from RNA to proteins.

RNA polymerase. An enzyme that forms RNA by copying the sequence of DNA within a gene. The profile of genes that actively form RNA within each cell is dictated by the selective recruitment of RNA polymerase to distinct genes.

scaling. An adjustment of pattern to size achieved by altering the distribution of the morphogen gradient that determines the pattern.

stem cell. A nondifferentiated cell that maintains the capacity to divide. Typically, at every division of a stem cell, one cell will retain the nondifferentiated features while the other will undergo differentiation to provide the supply of new cells to the tissue.

stem-cell niche. A physical compartment that maintains the stem cells in their nondifferentiated and proliferative form. Because the physical space within the niche is limited, only some of the progeny will remain as stem cells. The cells forced to leave the niche will undergo differentiation.

symmetry breaking. A local event that disrupts symmetry at the cell or organism level. Such events give rise to the early definition of the embryonic axes.

transcription. Synthesis of RNA from the DNA template of the gene; this takes place in the nucleus of the cell.

translation. Synthesis of proteins by the ribosomes based on the information provided by RNA molecules; this takes place in the cytoplasm of the cell.

Glossary of Scientific Images

This glossary provides additional information on the biological images used and the techniques employed to generate and acquire the images.

Trees of embryonic neurons. The sensory neurons of the peripheral nervous system of a 14-hour-old fruit fly embryo connect to the central nervous system, seen as a ladder at the bottom. They transmit information from external sense organs and internal muscle stretch receptors. The structure of the neurons is repeated accurately at every segment of the embryo. A process of "whole-mount" staining is used to visualize the global organization of the tissues. The specimen is fixed with a solution that generates chemical links between adjacent proteins, converting the soft tissue into a solid matrix. In parallel, the membranes

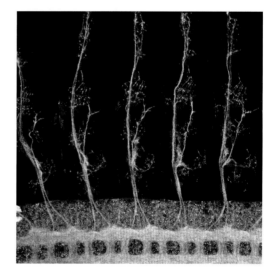

that separate the cells are dissolved, such that all parts of the specimen, including the interior ones, are now accessible to molecules that are introduced from outside. The specimen becomes a "sponge" that can be bathed in different solutions. Antibodies recognizing a specific protein expressed in some neural cell nuclei (red) or a protein associated with nerve membranes (gray) are used to mark distinct aspects of the nervous system. Finally, for optimal visualization by a fluorescence microscope, the embryo (measuring 0.5 × 0.2 mm) is sliced with a sharp tungsten needle along its back side and flattened, so that the entire structure of the nervous system can be visualized in one optical slice. *Page ii.*

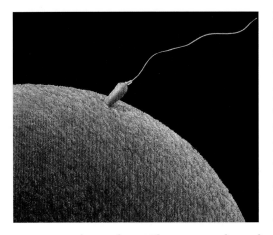

Human sperm and egg at the moment of penetration. The head of the sperm, which contains the nucleus, is buried in the follicular cells that surround the egg. The sperm must penetrate a membrane covering the egg in order to reach the female nucleus and initiate fusion of the male and female nuclei. Note the dramatic size difference between the sperm and the egg. The image was produced by a scanning electron microscope. The sample is coated with conductive materials such as gold and is targeted by an electron beam. The energy exchange between the electron beam and the sample results in a reflection of high-energy electrons, which can be traced by specialized detectors. This method allows high-resolution imaging of the outer surface of objects. The images are produced in black and white. The colors in this photo were added after acquisition. *Page x.*

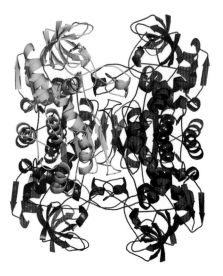

Diagram of protein structure. The backbone structure of the enzyme alcohol dehydrogenase that was purified from the bacterium *Thermoanaerobium brockii.* The active enzyme is composed of four identical protein chains (displayed in different colors) that combine to form the active enzyme. Both the assembly of the units and the recognition of the substrate require a specific and reproducible three-dimensional structure. This type of display is termed a ribbon diagram because it shows only the central backbone of the protein. In reality, the atoms and side chains that are connected to the backbone generate a structure that fills the space around the backbone. *Page 13.*

Skeletal muscle of mouse. Thousands of repeated units, called sarcomeres, are present within a muscle cell. Each sarcomere contains thin filaments (composed of actin polymers) at both edges, and thick filaments (composed of myosin proteins) at the center. The coordinated sliding of filaments in all sarcomeres generates the contraction of the muscle. A muscle cell specifically expresses the RNAs and proteins that are necessary to build the sarcomeric unit. Although this image was visualized with a scanning electron microscope, it does not show the outer surface of the muscle. Rather, the muscle was embedded in a resin that hardened. The resin was then broken, such that the surface now depicts the inner part of the muscle. *Page 14.*

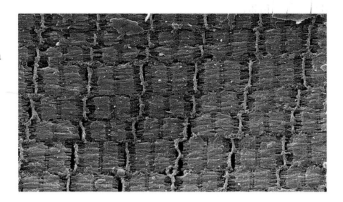

Mouse limb. The mouse limb contains an intricate array of muscles that are connected to the bones by tendons. Muscle contraction is triggered by nerves. A process of "whole-mount" staining is used to visualize the global organization of muscle, nerve, and tendon tissues. Because each tissue expresses a different set of proteins, these proteins can be used to mark the position and structure of each of the tissues. In this case, the specimen was incubated with three antibodies, each recognizing a distinct protein in muscle (red), nerves (blue), or tendons (green). Each of these antibodies is tagged with a fluorescent probe that displays a different color. When imaged under a microscope that can simultaneously detect all three fluorescent colors, each of the tissues is distinctly outlined. *Page 16.*

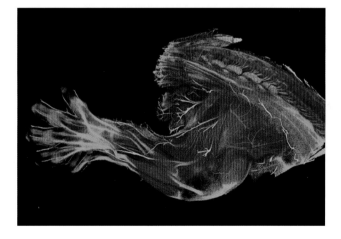

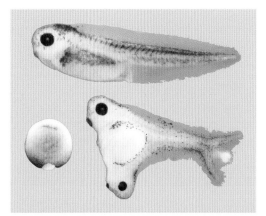

Spemann-Mangold transplantation experiment. In the classic 1924 Spemann and Mangold experiment, cells that will give rise to the head and trunk of a newt embryo about ten hours after fertilization were transplanted into a host embryo of a comparable or younger stage (bottom, left). This manipulation induced the cells of the host to form a Siamese twin within two days after the manipulation (bottom, right). The normal embryo (top) is 4 mm long. *Page 21.*

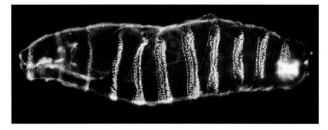

Outer "shell" of a fruit fly embryo. It takes 24 hours from fertilization until the fruit fly embryo hatches as a larva that contains 50,000 cells. Toward the end of embryogenesis, the outer cell layer secretes a hard shell called the cuticle, which protects the larva from drying and from mechanical insults. The cuticle is patterned according to the cells that secreted it; it displays the outlines of segments and a distinct polarity. In mutant embryos, defects in the structure of the cuticle imply a parallel flaw in the tissues that secreted it. Because the cuticle acts like an external skeleton and is highly resilient, it was technically much easier for scientists to identify defects in the cuticle than in the embryo itself, which has disintegrated. To visualize the cuticle, the internal tissues of the embryo are dissolved and the imaging is carried out with a light microscope using "dark field microscopy," which outlines the structures more distinctly. The size of the embryo is 0.5 × 0.2 mm. *Page 31.*

Developing eye of the fruit fly. The compound eye of the adult fly contains 800 repeated units called ommatidia, each of which contains: eight neurons that sense light (photoreceptors), cells that make up the lens, and pigmented cells that insulate each ommatidium. The formation of this elaborate organ starts at the larval phase in a tissue that is initially unpatterned. In a highly organized spatial and temporal manner, a new row of ommatidia is progressively generated every two hours. The global pattern is defined as regular spacing between ommatidia by inducing the formation of a "founder" photoreceptor cell for each ommatidium. The regular spacing between founder cells is determined by the local interactions among cells. In this way, the neighboring cells are inhibited from assuming a similar fate. After the position and fate of the founder cell is established, it instructs its immediate neighbors to form the other photoreceptors of the ommatidium. The process of whole-mount staining uses antibodies that bind with and mark a nuclear protein of the founder cells, shown here in blue. *Page 36.*

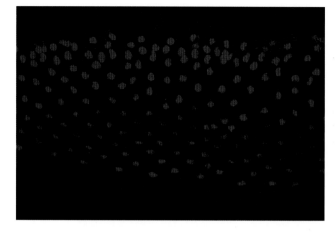

Developing sense organs of the fruit fly thorax. Hairs that cover the thorax of the fly serve as sense organs that detect mechanical and chemical stimuli. These hairs are positioned at regular intervals that are determined during pupariation (the pro-

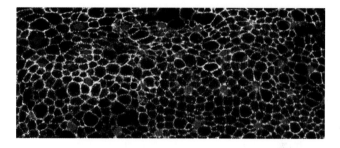

cess that dissolves the larval tissues and assembles the future tissues of the adult). The image shows the future thorax in a pupa, wherein one antibody detects the membranes of the cells (white) while the other antibody detects a protein that is specifically expressed in the nuclei of future sense organs (red). The regular spacing between the sense organs is achieved through cell communication, which locally inhibits neighboring cells from assuming the same fate. *Page 38.*

Monitoring mutant cells. The role of cell communication in establishing the regular distribution of sense organs in the fruit fly pupa can be unraveled when we follow the consequences of obliterating this communication. Eliminating the receptor that receives the signals from neighboring cells makes the cell "deaf" to these cues. However, because the same receptor is required for earlier processes of development in the embryo, development will be arrested at embryogenesis if it is completely re-

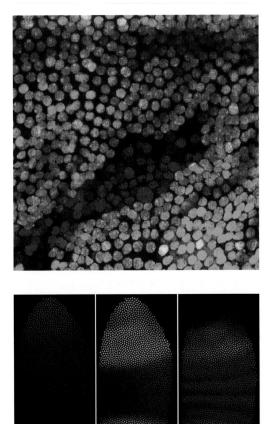

moved. To overcome this problem, scientists use genetic tricks to eliminate the receptor in a small group of cells in the pupa while leaving most of the cells intact. Green marks the normal cells, so the absence of green indicates the mutant "clone." Because all cells have the potential to become sense organs, when the communication among cells is absent, the entire clone of mutant cells assumes the fate of the sense organ (red). The neighboring cells of the same tissue that contain the receptor display the normal spacing of sense organs. *Page 39.*

Generating segments in the fruit fly embryo. A pattern is formed by a series of successive decisions that increase the complexity and refine the pattern. This image shows three stages in the generation of a pattern along the head-tail axis (from left to right, the future head is at the top). In these stages, which take place two to three hours after fertilization, the embryo contains 5,000 nuclei that are positioned along its periphery. First, a protein is distributed in a gradient that displays maximal levels at the head region (blue). This gradient, which represents the first symmetry-breaking event and triggers the patterning process, instructs the expression of target genes in distinct parts of the embryo. For example, the protein marked in green is expressed only in regions of high or low levels of the gradient, with sharp boundaries. Communication driven by proteins that are expressed in distinct zones of the embryo leads to the expression of the next cohort of genes. The protein shown in red is expressed in a repeated pattern of seven stripes, which will serve to define the 14 segments of the embryo. *Page 45.*

Convergence of signaling pathways. This image shows distinct groups of cells that were defined in a five-hour-old fruit fly embryo. Within every segment, each group is defined by a different signaling pathway and expresses distinct sets of proteins. Representative proteins are shown here in red and blue. In the next step, a new cell fate that requires simultaneous activation of the two signaling pathways will be induced. This will take place only in the cells in which both signals converge. *Page 48.*

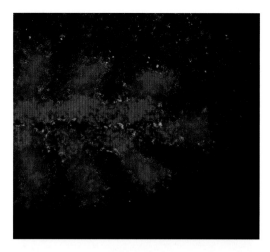

Responses of cells reflect their history. Cells of a quail embryo were grafted in a duck embryo. The quail cells are shown in black in the left panel and were marked using antibodies that recognize a quail protein. A gene that is present in both quail and duck genomes is expressed only in the quail cells (displayed as white spots on the right panel). Thus, even

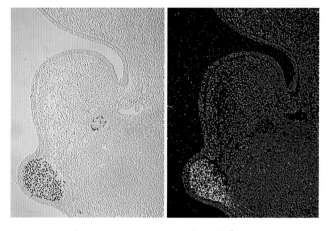

though the duck and quail cells are present in the same environment, their different histories lead to distinct responses. *Page 52.*

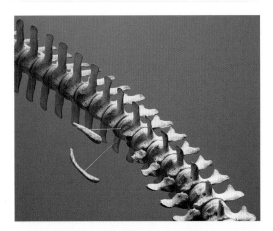

Vertebrae represent repeated units with subtle modifications. The skeleton of a killer whale shows the modular nature of the vertebrae. We can also see the specific alterations that are introduced along the head-tail axis to comply with the distinct structure and function of body parts that are formed at different locations. *Page 53.*

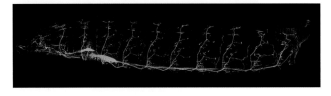

Segments in the fruit fly larva display a repeated unit with subtle modifications. The body of invertebrates is divided into segments. In the larva of the fruit fly, a visualization of the nerves (green) and the body wall muscles they associate with (blue) shows a segmentally repeated structure. A closer look identifies subtle modifications along the different segments, to comply with the distinct functions of each segment. *Page 54.*

Patterning over long distances by morphogens. This image shows the cells in the fruit fly larva that will give rise to the adult wing. At this stage, the cells form a single layer. The information on the global position of each cell is provided by a protein that is expressed by a small group of cells (white, on the left side) and secreted to the extracellular milieu. The protein's diffusion to neighboring cells generates a gradient of decreasing concentration based on the distance from the morphogen source. According to the level of morphogen each cell encounters, the cell can assess its position. This tissue preparation shows the morphogen (white) in a spotted form inside the cells that received the signal, after it has been captured and incorporated into specialized organelles. The level of morphogen inside the cells is presumed to reflect the corresponding quantities outside the cell. *Page 61.*

Scaling pattern with size. When flies are grown under altered nutritional conditions, they grow to different sizes but maintain their relative proportions. This image shows the patterning of the wing by the stereotyped positions of the veins. Scaling causes the relative position of veins to be fixed regardless of the total wing size. *Page 63.*

Reading gradients over time. This image shows a ten-day-old mouse embryo, in which the future fore- and hind limbs can be seen as petals. A morphogen termed Sonic hedgehog is expressed only in the very posterior part of the limb (the future "pinkie"). The protein is secreted outside the cell and dispersed along the entire limb bud. In this image, the RNA of *sonic hedgehog* is labeled (purple) within the cells, specifically marking the small group of cells in which it is produced. A similar protocol to whole-

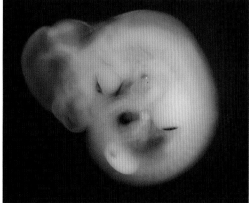

mount protein labeling was used, but in this case the specific RNA was identified by tagged DNA or RNA sequences that bind to the sonic hedgehog RNA. As a result, cells that express this specific RNA molecule can be identified within the context of the entire organism. These cells are visualized by an enzyme that deposits dye only in the cells where it associates with the tagged probe that detected the *sonic hedgehog* RNA. *Page 69.*

Evolution of patterns. This image depicts the wings of fly species that diverged over the past 60 million years. The strikingly different designs of pigmentation are derived from alterations in the pattern generated by enzymes that produce the pigments. Although the patterns differ greatly, the structure of the enzymes themselves has not been modified over this period. This indicates that mutations that alter the expression of key genes can lead to dramatic changes in pattern while leaving unaltered the protein machinery that executes the process. *Page 76.*

Predictable cell lineage in the worm. The embryo of the worm displays a highly stereotyped developmental program in which the position and fate of every cell is invariant. The orientation of cell division, as well as asymmetric segregation of proteins between daughter cells, is coordinated based on the global position of the cell within the embryo. The images represent snapshots of a movie of the *same* embryo, in which specific structures termed P granules are marked by the expression of a green fluorescent protein that can be visualized in live cells and embryos. Future head structures are to the left. In the two-cell stage (upper left), P granules are concentrated in the more posterior cell. Upon each division, these granules continue to be further concentrated only within a single cell. The cell that will eventually contain these granules will give rise to the tissue producing sperm or egg. *Page 83.*

Generating reproducible patterns.
These images depict the region connecting the wing to the body in six different fruit flies of the same species. When we follow the hair structures formed, every minute detail is identical, including the number, position, and type of each hair. Such patterning can be

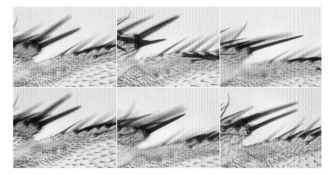

achieved accurately by cell communication because intricate regulatory and feedback mechanisms operate to keep all processes in check. *Page 88.*

Vestibular hair cell. The cells shown here are located in the inner ear of the mouse. Each cell displays multiple hairlike structures that project to the fluid-filled canal. The direction of these hairs provides the organism with a sense of orientation in three-dimensional space, and hence with balance. The hairs marked by an artificial coloring emanate from a single cell. The image was acquired by a scanning electron microscope. *Page 102.*

Formation of the tracheal tubular network. This image shows an interconnected network of tubes that delivers air directly to the different organs of the insect. This network is formed from clusters of designated cells in each segment, there are a total of 800 cells in this tissue on each side of the embryo. During embryogenesis these cells migrate in a stereotypic fashion, following cues provided by their environment. The final

pattern of the tissue reflects the history of migration of these cells. In the image, the cells forming the trachea were visualized in a 12-hour-old fruit fly embryo through the detection of a protein (brown) specifically expressed in this tissue. *Page 106.*

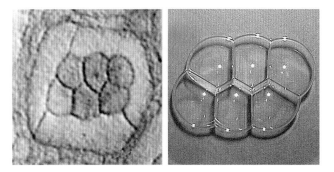

Formation of lens cells in the compound eye. The compound eye of the adult fly is generated during the larval and pupal stages. The cells that will form the lens of each ommatidium are adjacent to each other (left image). The tissue was incubated with a stain that marks the membranes of the cells and thus allows the visualization of their outlines. The arrangement of these cells is dictated by the physical surface tension of their adjoining membranes. Thus, it is possible to generate an identical arrangement by a similar number of adjoining soap bubbles (right image). *Page 108.*

Generating compartments. Distinct domains within the developing organism maintain their integrity as distinct compartments and minimize contact with the adjacent ones. This process is executed by the expression of specific cell surface proteins that promote association between cells expressing the same set of proteins. Indicated by fluorescent green, the reporter protein shows the expression pattern of a gene that defines the posterior compartment of each segment in the fruit fly. Visualization was done under a fluorescence-dissecting microscope on a live fly. The sharp borders and lack of mixing with the adjacent compartments are notable. *Page 110.*

Respecting compartment boundaries. In the larval tissue that will give rise to the adult fruit fly wing, cells that descended from a single progenitor cell are marked in red. Although they mix freely with the cells in their own compartment on the left, they minimize contact with the cells in the adjacent compartment and generate a sharp border. The image was achieved by whole-mount staining for a protein in red that is expressed only in the cells that descend from the original founder cell. *Page 112.*

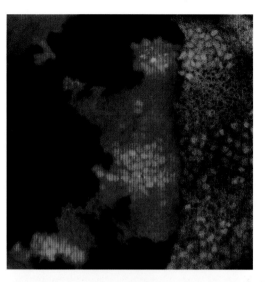

Defining global orientation in the developing wing. Each cell in the developing pupal wing of the fruit fly forms a single hair. The hairs of all cells point in the same direction. The tissue was probed for the polymeric form of actin, a major constituent of the hairs. Because the actin accumulates below the membrane, it also marks the outline of each cell. The uniform orientation of the hairs is achieved by communication between adjacent cells that amplifies a response to a global orientation signal. *Page 115.*

Selective killing of cells defines the digits. To generate the digits, cells from the interdigit zones are removed, as shown in this image of a limb from a 13-day-old mouse embryo. The regions of active cell death were identified by using a dye that specifically marks lysosomes, the cellular organelles where proteins are degraded. Active cell killing is achieved by a process termed autophagy, in which cells devour themselves from within. *Page 117.*

The stem cell niche. The lens of the fish eye is surrounded by a physical compartment that defines a niche for stem cells. This niche provides the stem cells with the proteins that maintain them in the proliferative, nondifferentiated state. Once the cells leave the niche, they differentiate to produce neurons (black). The physical restrictions of the niche ensure that the number of stem cells will be fixed throughout the life of the organism. This image was acquired by whole-mount antibody staining. *Page 122.*

Induced pluripotent stem cells. Recent technologies enable scientists to revert differentiated cells back to their nondifferentiated state. From there, the cells can be directed to differentiate in new ways. This provides the capacity to generate new organs, without relying on the endogenous source of stem cells. Human skin cells were reverted to their pluripotent state, and then directed to differentiate in culture as neural cells. Cells that are actively dividing as stem cells can be identified by the expression of proteins that are elevated during cell division (pink). Their progeny show a rosette structure that is typical of nerve cells in culture. *Page 130.*

Tissue regeneration. The planarian flatworm exhibits a remarkable capacity to regenerate. A small part of a dissected animal can give rise to a whole worm. In this image, cuts were made at different places along the body of the worm. A new head was formed at each

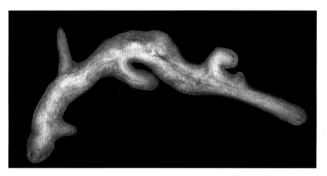

of these places, giving rise to a six-headed worm. *Page 132.*

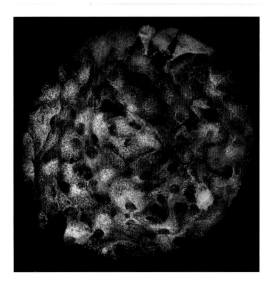

The insulin factory. Insulin is formed in the pancreas in thousands of cell clusters, or "balls," called the islets of Langerhans. Each aggregate contains several cell types, including the beta cells that produce insulin. The tiny dotted white staining shows the insulin, stored in vesicles within these cells. When blood sugar levels are elevated, beta cells are triggered to release insulin into the bloodstream. Juvenile diabetes is caused by the pathological death and elimination of the islets of Langerhans. The red and green staining shows distinct proteins that were artificially expressed in these cells for an experiment. For our purpose they help to outline the different cells. *Page 137.*

Recommended Reading and Web Resources

Chapter 1: Introduction

Hooke, R. *Micrographia: or, Some Physiological Descriptions of Minute Bodies Made by Magnifying Glasses*. London, 1665. Turning the Pages Online Project, US National Library of Medicine. http://archive.nlm.nih.gov/proj/ttp/flash/hooke/hooke.html.

Shilo, B. Z. Conveying principles of embryonic development by metaphors from daily life. *Development (Spotlight)* 140 (2013): 4827–4829. http://dev.biologists.org/content/140/24/4827.long/.

Shilo, B. Z. Egg to organism: Visualizing the concepts of development. *EMBL Science and Society Seminars* (2012). http://medias01-web.embl.de/Mediasite/Play/102f090bfa 93473eb369a8c646a8d09c1d?_ga=1.25625853.1160088470.1393225479.

Shilo, B. Z., and Weinberg, R. A. DNA sequences homologous to vertebrate oncogenes are conserved in Drosophila melanogaster. *Proceedings of the National Academy of Sciences USA* 78 (1981): 6789–6792.

Weinberg, R. A. A molecular basis of cancer. *Scientific American* 249 (1983): 126–142.

Chapter 2: It's All in the Genes

Deciphering the genetic code: Marshall Nirenberg: Glossary. Office of NIH History. http://history.nih.gov/exhibits/nirenberg/glossary.htm.

Ramakrishnan, V. 2009 Nobel lecture. http://www.nobelprize.org/nobel_prizes /chemistry/laureates/2009/ramakrishnan-lecture.html.

Tjian, R. Molecular machines that control genes. *Scientific American* 272 (1995): 54–61.

Transcription: why genes are on only when you need them. modENCODE project, AAAS. http://modencode.sciencemag.org/transcription/introduction/.

Yonath, A. E. 2009 Nobel lecture. http://www.nobelprize.org/nobel_prizes/chemistry/laureates/2009/yonath-lecture.html.

Chapter 3: How an Embryonic Cell "Decides" Its Future Destiny

De Robertis, E. M. Spemann's organizer and the self-regulation of embryonic fields. *Mechanisms of Development* 126 (2009): 925–941.

Horder, T. History of developmental biology. In *Encyclopedia of Life Sciences.* Chichester: John Wiley and Sons, 2010.

Chapter 4: How Cells Talk and Listen to Each Other

Drosophila as a model organism: an introduction. modENCODE project, AAAS. http://modencode.sciencemag.org/drosophila/introduction/.

Nüsslein-Volhard, C. 1995 Nobel lecture. http://www.nobelprize.org/nobel_prizes/medicine/laureates/1995/nusslein-volhard-lecture.pdf.

Pellettieri, J. Online developmental biology: introduction to *Drosophila.* https://www.youtube.com/watch?v=ePBghFrPb7Y.

Wieschaus, E. H. 1995 Nobel lecture. http://www.nobelprize.org/nobel_prizes/medicine/laureates/1995/wieschaus-lecture.pdf.

Chapter 5: How Do Simple Modules Lead to Complex Patterns?

Artavanis-Tsakonas, S., and Muskavitch, M. A. Notch: the past, the present, and the future. *Current Topics in Developmental Biology* 92 (2010): 1–29.

Grimm, O., Coppey, M., and Wieschaus, E. 2010. Modelling the Bicoid gradient. *Development* 137 (2010): 2253–2264.

Kmita, M., and Duboule, D. Organizing axes in time and space; 25 years of collinear tinkering. *Science* 301 (2003): 331–333.

McGinnis, W., and Kuziora, M. The molecular architects of body design. *Scientific American* 270 (1994): 58–66.

Perrimon, N., Pitsouli, C., and Shilo, B.-Z. Signaling mechanisms controlling cell fate and embryonic patterning. *Cold Spring Harbor Perspectives in Biology* 4 (2012): a005975.

Pourquie, O. Vertebrate segmentation: from cyclic gene networks to scoliosis. *Cell* 145 (2011): 650–663.

St. Johnston, D., and Nüsslein-Volhard, C. The origin of pattern and polarity in the *Drosophila* embryo. *Cell* 68 (1992): 201–219.

Chapter 6: How Can a Single Substance Generate Multiple Responses?

Ben

Ben-Zvi, D., Shilo, B.-Z., and Barkai, N. Scaling of morphogen gradients. *Current Opinions in Genetics and Development* 21 (2011): 704–710.

Nüsslein-Volhard, C. Gradients that organize embryo development. *Scientific American* 275 (1996): 54–61.

Riddle, R. D., and Tabin, C. J. How limbs develop. *Scientific American* 280 (1999): 74–79.

Rogers, K. W., and Schier, A. F. Morphogen gradients: from generation to interpretation. *Annual Reviews in Cell and Developmental Biology* 27 (2011): 377–407.

Chapter 7: How Do Patterns Evolve Rapidly?

Carroll, S. B. Evo-devo and an expanding evolutionary synthesis: a genetic theory of morphological evolution. *Cell* 134 (2008): 25–36.

Carroll, S. B., Prud'homme, B., and Gompel, N. Regulating evolution. *Scientific American* 298 (2008): 60–67.

Chapter 8: How Are Cells Programmed to Follow Predictable Routes to Specification?

C. elegans as a model organism: an introduction. modENCODE project, AAAS. http://modencode.sciencemag.org/worm/introduction/.

Li, R. The art of choreographing asymmetric cell division. *Developmental Cell* 25 (2013): 439–450.

Lyczak, R., Gomes, J. E., and Bowerman, B. Heads or tails: cell polarity and axis formation in the early *Caenorhabditis elegans* embryo. *Developmental Cell* 3 (2002): 157–166.

Pellettieri, J. Online developmental biology: introduction to *C. elegans*. https://www.youtube.com/watch?v=zc1P7lGSzdU.

Prehoda, K. E. Polarization of *Drosophila* neuroblasts during asymmetric division. *Cold Spring Harbor Perspectives in Biology* 1 (2009): a001388.

Sulston, J. E. 2002 Nobel lecture. http://www.nobelprize.org/nobel_prizes/medicine/laureates/2002/sulston-lecture.html.

Updike, D., and Strome, S. P granule assembly and function in *Caenorhabditis elegans* germ cells. *Journal of Andrology* 31 (2010): 53–60.

Chapter 9: Ensuring That Embryo Development Is on the Right Track

Horvitz, H. R. 2002 Nobel lecture. http://www.nobelprize.org/nobel_prizes/medicine/laureates/2002/horvitz-lecture.html.

Lander, A. D. How cells know where they are. *Science* 339 (2013): 923–927.

Perrimon, N., and McMahon, A. Negative feedback mechanisms and their roles during pattern formation. *Cell* 97 (1999): 13–16.

Chapter 10: Shaping the Tissues

Dahmann, C., Oates, A. C., and Brand, M. Boundary formation and maintenance in tissue development. *Nature Reviews in Genetics* 12 (2011): 43–55.

Hayman, A. A., and Simons, K. Beyond oil and water: phase transitions in cells. *Science* 337 (2012): 1047–1049.

Paluch, E., and Heisenberg, C.-P. Biology and physics of cell shape changes in development. *Current Biology* 19 (2009): R790–799.

Rørth, P. Fellow travellers: emergent properties of collective cell migration. *EMBO Reports* 13 (2012): 984–991.

Chapter 11: Stem Cells

Hochedlinger, K. Your inner healers. *Scientific American* 302 (2010): 46–53.

Inoue, H., Nagata, N., Kurokawa, H., and Yamanaka, S. iPS cells: a game changer for future medicine. *EMBO Journal* 33 (2014): 409–417.

Morrison, S. J., and Spradling, A. C. Stem cells and niches: mechanisms that promote stem cell maintenance throughout life. *Cell* 132 (2008): 598–611.

Muneoka, K., Han, M., and Gardiner, D. M. Regrowing human limbs. *Scientific American* 298 (2008): 56–63.

Pedersen, R. A. Embryonic stem cells for medicine. *Scientific American* 280 (1999): 68–73.

Tanaka, E. M., and Reddien, P. W. The cellular basis for animal regeneration. *Developmental Cell* 21 (2011): 172–185.

What is a stem cell? What are embryonic stem cells? What are induced pluripotent stem cells? http://www.youtube.com/watch?v=_hbgeQzmU9U&feature=youtu.be.

Chapter 12: What's Next

Ackerman, J. The ultimate social network. *Scientific American* 306 (2012): 36–43.

Capecchi, M. R. Targeted gene replacement. *Scientific American* 270 (1994): 52–59.

Gage, F. H. Brain, repair yourself. *Scientific American* 289 (2003): 46–53.

Hall, S. S. Diseases in a dish. *Scientific American* 304 (2011): 40–45.

Little, S. C., and Wieschaus, E. F. Shifting patterns: merging molecules, morphogens, motility, and methodology. *Developmental Cell* 21 (2011): 2–4.

Sasai, Y. Grow your own eye. *Scientific American* 307 (2012): 44–49.

Index

Page numbers in *italics* refer to illustrations.